Exploring the SOUTH LAND

Tasmania emerges from
Terra Australis Incognita

Libby and John McMahon

MARITIME MUSEUM OF TASMANIA

ISBN: 0-9757237-2-3

Published in 2006 for the Maritime Museum of Tasmania Inc. by 40 Degrees South Pty Ltd

Editors: Lyn Wilson and Warren Boyles
Book Design: Kent Whitmore

Tasmania   HOBART CITY COUNCIL

The Maritime Museum of Tasmania gratefully acknowledges the assistance of the Department of Tourism, Arts and the Environment, the Tasmanian Museum and Art Gallery and the Hobart City Council.

FRONT COVER: Part of *Magnum Mare del Zur cum Insula California* 1680 de Wit

BACK COVER: Part of *Mar di India* c. 1700 Jansson

NOTE: The original spelling of each map title has been retained in the text.

Table of Contents

Foreword

The publication of *Exploring the South Land* by the Maritime Museum of Tasmania and *Tasmania 40°South* coincides with the 400th anniversary of the earliest documented arrival of navigators from beyond our shores: in 1606 Captain Willem Janszoon in the vessel *Duyfken* sailed into Australian waters charting 300 kilometres of its coast, the first recorded European to do so.

Thousands of years before these explorers arrived, Aborigines had walked across the land bridge from mainland Australia to what was to become their island home (later to be called Van Diemen's Land, then Tasmania). In the 17th century European navigators began to chart the land they called *Terra Australis Incognita* and, as discoveries were made, Tasmania emerged. Maps from the unique Lamprell Collection are reproduced in *Exploring the South Land* to document these discoveries.

Dr Lamprell, who bequeathed the collection to the Tasmanian Museum and Art Gallery and the Maritime Museum of Tasmania, was born and educated in London. He became a medical officer in India in 1931 and during the war reached the rank of lieutenant colonel. After retiring from medical practice in Malaya in 1967, Dr Lamprell and his family settled in Tasmania. It was then that he assembled his map collection and was subsequently appointed a Fellow of the Royal Geographic Society, London.

The Tasmanian Government, through the Department of Tourism, Arts and the Environment, contributed to the project, 'Australia on the Map 1606—2006', making possible the task of conserving the Lamprell Collection for publication and exhibition. In addition, the Tasmanian Museum and Art Gallery provided two more maps for reproduction and Philip Fowler supplied one map from his collection.

We are grateful to the authors, Libby and John McMahon, for undertaking the task of researching and documenting the Lamprell Collection, and to the *Tasmania 40°South* crew: Lyn Wilson, Warren Boyles and Kent Whitmore. The authors were assisted by Jo Huxley of the Tasmanian Museum and Art Gallery and Rona Hollingsworth, the Maritime Museum's curator. We also thank Associate Professor Peter Davis, of the University of Tasmania, for the Latin translations, and members of the *Alliance Française* who helped translate the text from the early French maps.

Colin Denny, President
Maritime Museum of Tasmania Inc.

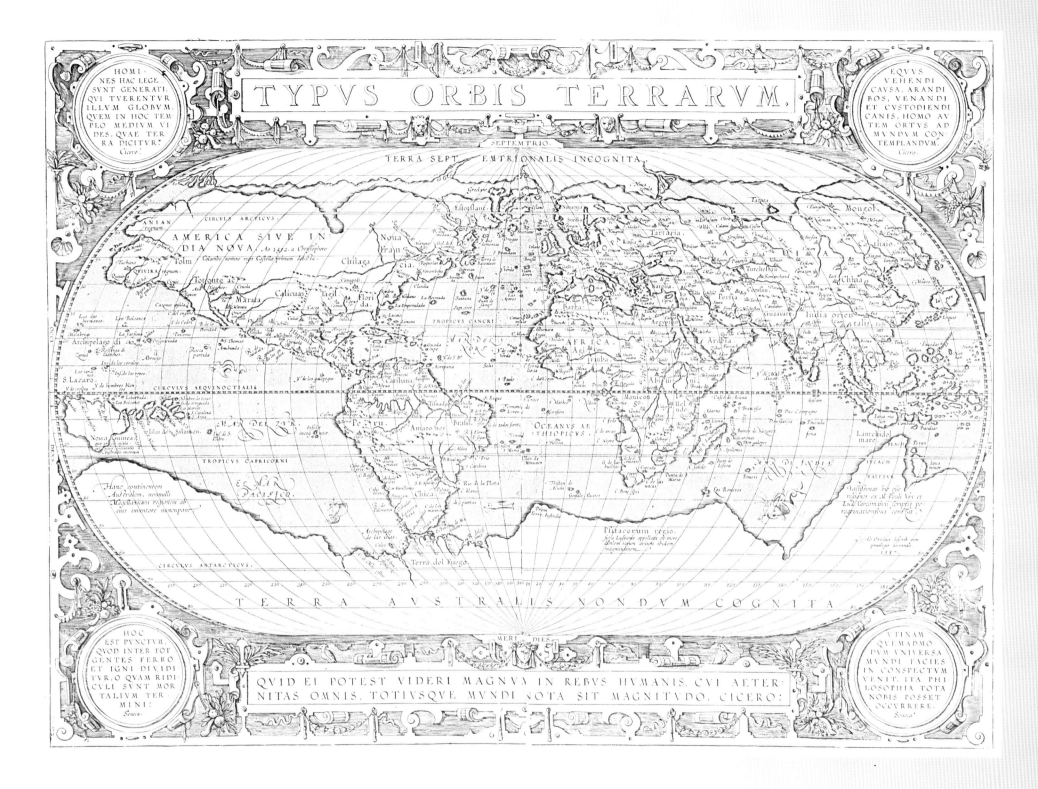

Introduction

In the 2nd century AD, the great cartographer Claudius Ptolemy produced his *Geographica*, which gave co-ordinates of latitude and longitude for thousands of features worldwide and included detailed instructions for creating maps. More than 1200 years later, in the 15th century, Ptolemy's *Geographica* was translated into Latin and the first printed maps were produced in Europe, giving contemporary Europeans the concept of a vast unknown 'South Land' — *Terra Australis Incognita*. This idea was not new, and in ancient times Ptolemy himself believed the world to be round, with Europe and Asia 'at the top'. The existence of a large hypothetical southern land was also deemed necessary to balance this great ball.

By the 17th century, with the rise in Dutch maritime power, Amsterdam was the cartographic centre of the world, with its mapmakers providing glimpses of the mysterious South Land. This intricate Dutch, and later French, cartography was enhanced by advances in printing techniques. The introduction of copperplate engraving (and in the early 19th century, steel plates) in which the design is 'drawn' or engraved onto a plate allowed much greater artistic freedom than the use of wood blocks, which, in contrast, required all areas apart from the design to be removed. Now flowing lines, delicate illustrations and 'watered silk' seas were possible. Printing plates were frequently updated and re-used, sometimes over a period of years. Engraved plates were also acquired by, or sold to, rival cartographers whose own publishing houses then printed the maps.

Until the beginning of the 19th century, and the invention of lithography, colouring was almost always applied by hand. Artists painted in watercolour onto the newly printed maps, which were then bound into an atlas, or sold individually. Some cartographers sold the same editions both coloured and uncoloured.

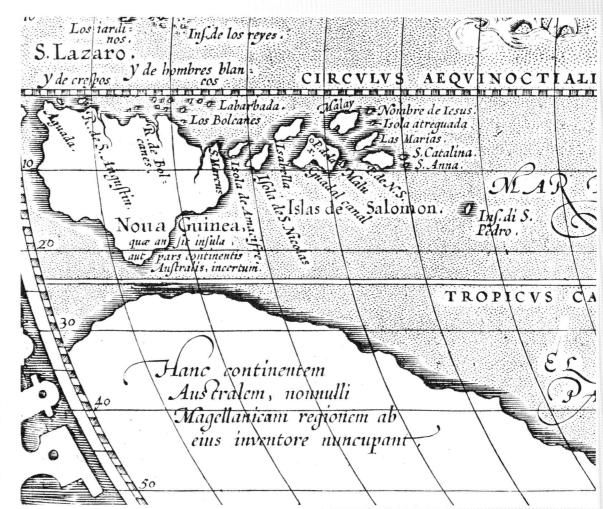

Typus Orbis Terrarum

REPRODUCTION, FIRST ISSUED **1587 Ortelius**

Detail

New Guinea is shown above the unknown South Land, which
has the Latin text 'Some call this continent Australis, some
the Magellanic region after its discoverer'.

Dutch Cartographers

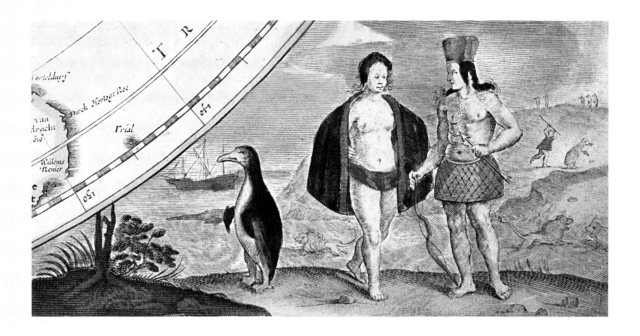

While copperplate engraving improved the visual appearance of maps, trading rivalry provided the impetus for exploration, which supplied new material to 17th century mapmakers. In 1602 the Dutch East India Company was established, and competition with the Portuguese in the East Indies became intense. Ships were dispatched to find faster routes and new trading opportunities, seeking lands rich in spices, gold and silver. Information brought back by these explorers was used by cartographers to make new maps, or was added to existing copperplates. However, some information was jealously guarded and kept secret to protect trading advantages.

In 1606, Captain Willem Janszoon (or Jansz), while on an exploratory voyage from the Dutch East Indies, made the first recorded European contact with Australia. Sailing the *Duyfken* east along the south coast of New Guinea and nearing Torres Strait, Janszoon was forced south by unfavourable conditions and dangerous shoals. He encountered the western side of Cape York near the Pennefather River. Continuing south along what he believed was an extension of the coast of New Guinea, to a point he named 'Cabo Keer-weer' (Cape Turnagain), he then retraced his route. Janszoon had traversed about 300 kilometres of the coast of Cape York but failed to find Torres Strait. 'Keerweer' is noted on some 17th century maps on the coast of New Guinea. The section of Cape York explored by Janszoon, with Cape Keer-weer shown, is indicated in an enlargement from J. Arrowsmith's 1848 map *Eastern Portion of Australia* (p. 55).

In the same year that Janszoon reached Cape York, Captain Luis de Torres sailed east to west through the strait that now bears his name, proving that New Guinea was not attached to Australia. As he was in the Spanish service, Torres's voyage was not disclosed to the Dutch, although they suspected the existence of such a strait. This is often indicated on 17th century Dutch maps by a break in the coastline between Cape York and New Guinea.

Polus Antarcticus
1657 **Hondius**[2]

First published in 1641 by Hondius. This issue, published by Johannes Jansson, is one of the first maps to record Tasman's 1642—1643 discoveries with Van Diemen's Land and New Zealand added to the plate.

Detail
This fine vignette includes one of the earliest depictions of a penguin.

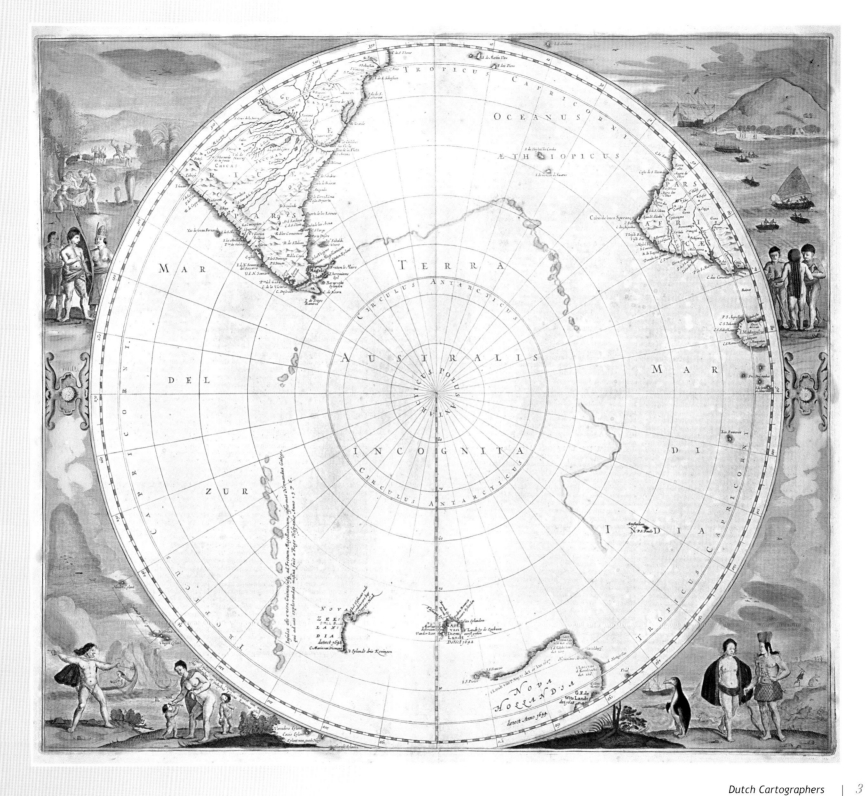

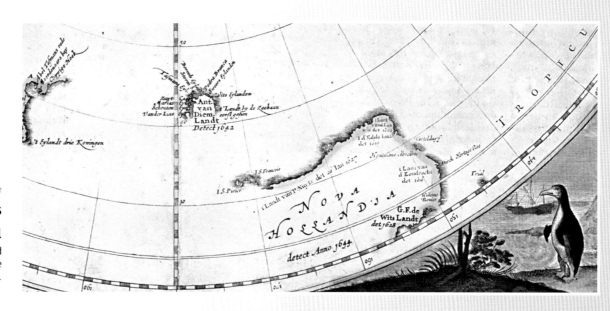

Polus Antarcticus

1657 **Hondius**

Detail

By the middle of the 17th century, the Dutch had encountered and mapped a recognisable portion of the coastline of Australia including Tasmania.

In 1642, Abel Janszoon Tasman (1603—1659) departed from Batavia (Jakarta) with two small ships, *Heemskerck* and *Zeehaen*. Having already charted parts of the west and south coasts of Australia, the Dutch hoped to find new areas for commercial exploitation and alternative routes to the Pacific. They also thought it possible that a separate landmass existed below 50° South. Therefore, Tasman was ordered to probe as far south as possible.

Sailing first to Mauritius, then into the Roaring Forties, Tasman set course to the east. Passing well south of Cape Leeuwin, the expedition encountered no land until 24 November 1642, when the west coast was sighted of what eventually was to be called Tasmania. Naming it Van Diemen's Land, Tasman proceeded to coast around the south of the island, exploring and roughly charting as he proceeded. Various features including Storm Bay and Maria Island were named. On 2 December, near North Bay on Forestier Peninsula, fresh water and edible greens were collected to supplement the shipboard diet. The following day Tasman decided to formally take possession of the area. However, with conditions unsafe to land a boat, the ship's carpenter swam ashore and planted the flag, claiming the land for the Dutch. Continuing to a point north of Schouten Island (which he named), Tasman then headed east to New Zealand. Tasman's crew are the first known Europeans to have set foot in Van Diemen's Land.

On a second voyage in 1644 one of Tasman's instructions was to establish whether or not Torres Strait existed. Sailing east along the south coast of New Guinea, confronted with numerous shoals as he neared Torres Strait, Tasman turned south. He subsequently explored and mapped the west coast of Cape York and the Gulf of Carpentaria, including the coast charted by Janszoon in 1606. He continued west across the north of Australia and down towards North West Cape in Western Australia before returning to Batavia.

To the Dutch East India Company, Tasman's expeditions were a disappointment. Apart from finding no new rich trading areas, he had failed to establish the eastern extent of Australia, or the existence of Torres Strait. Nor had he discovered a new landmass in the high southern latitudes. Despite these perceived failures, to prevent assisting trading rivals, only the company's mapmakers were given details of these expeditions. Tasman's notes and journals disappeared into the archives of the Dutch East India Company, and it was many years before information from his voyages was available to most cartographers.

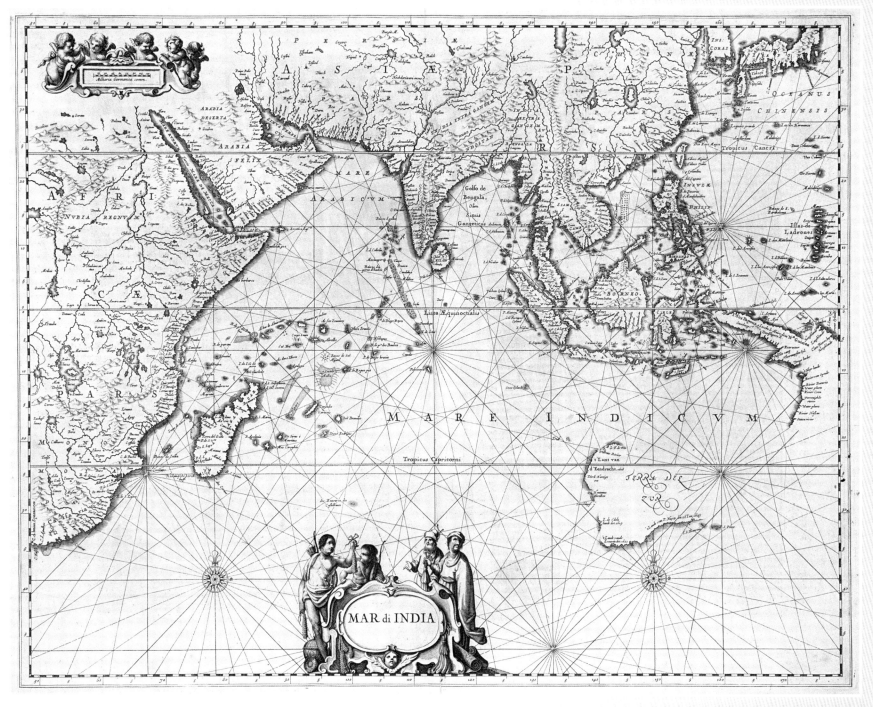

Mar di India

1650 Jansson[3]

Although this map was published in Jansson's atlas in 1650, and again unchanged in 1657, Tasman's discoveries are not shown.

A broken coastline indicates that the Dutch were aware of the possible existence of Torres Strait.

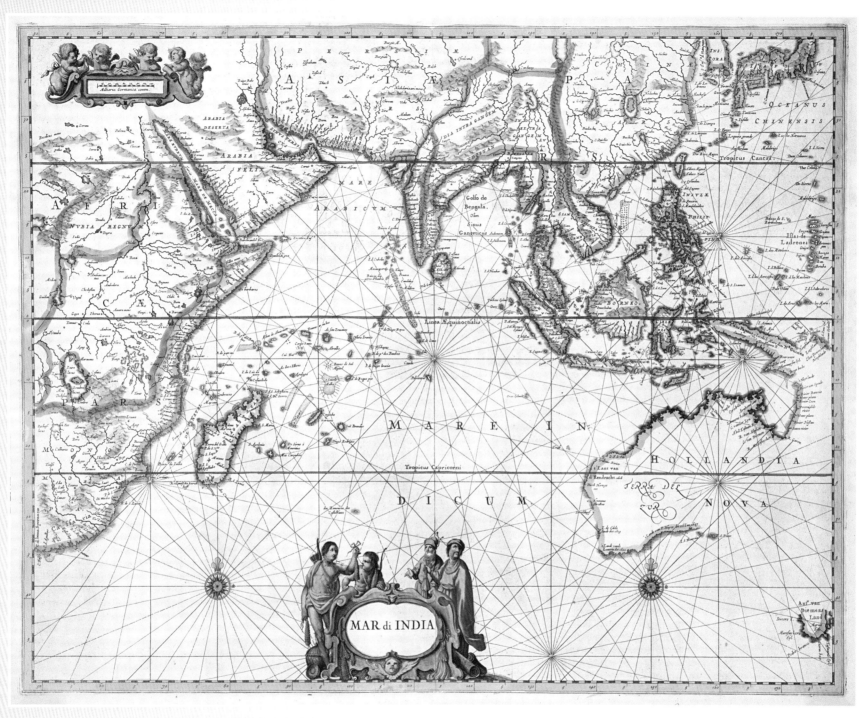

Mar di India

c. 1700 **Jansson**

In this issue, published after Jansson's death, information from Tasman's voyages has been added to the 1650 map and, apart from the uncharted east coast, a reasonably accurate outline of Australia and the south coast of Tasmania has emerged.

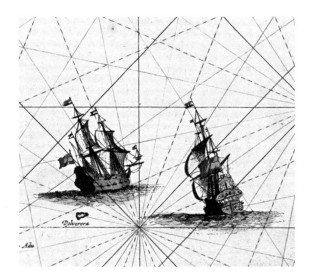

Paskaerte Zynde t'Oosterdeel Van Oost Indien

1666 Goos[4]

This decorative map with finely drawn ships was engraved, printed and published by Goos in his atlas *De Zee Atlas ofte Water-Waereld* in 1666. It records Tasman's 1644 voyage, but not his 1642–1643 expedition during which he encountered and named Van Diemen's Land.

Detail, right

In 1644, Tasman, having failed to penetrate Torres Strait, continued south and then west, exploring and mapping the Gulf of Carpentaria.

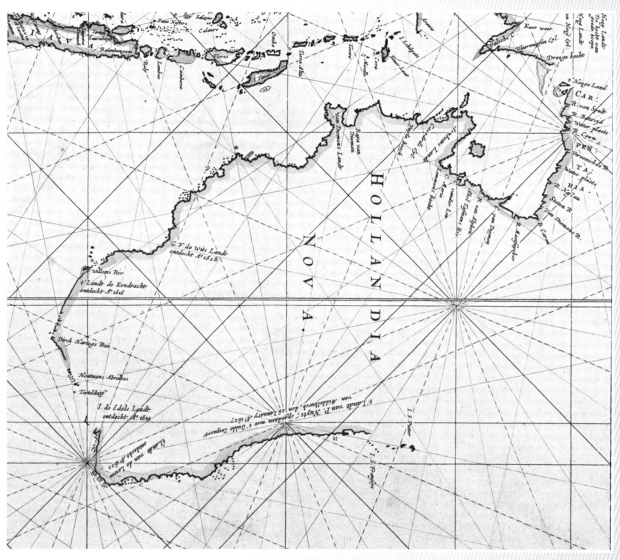

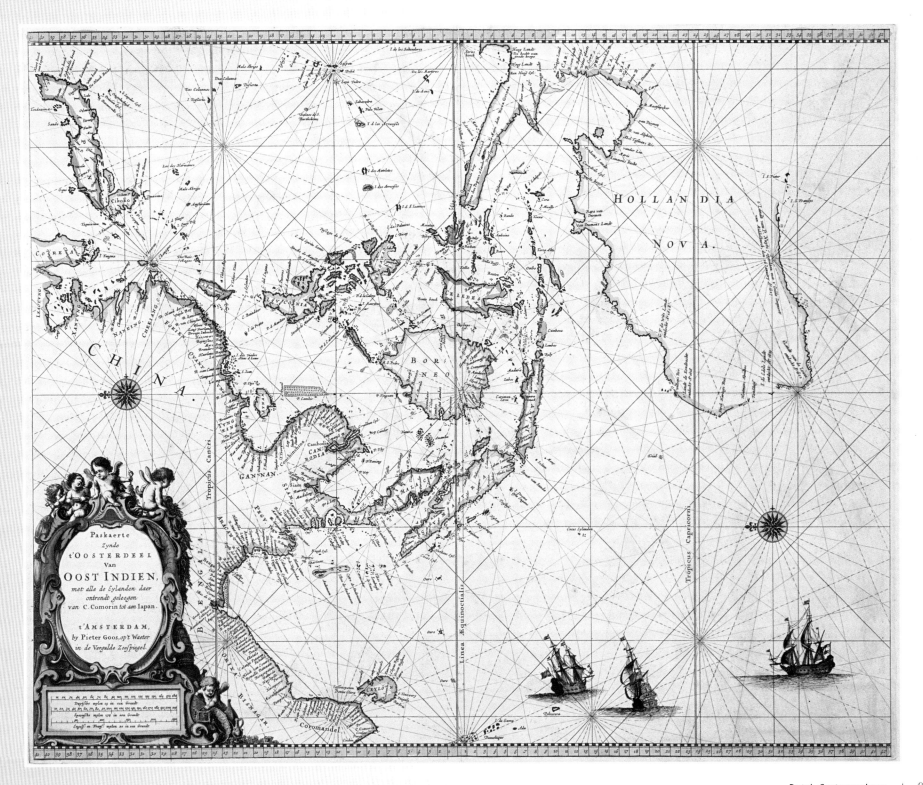

Magnum Mare del Zur cum Insula California

1680 de Wit[5]

Clearly recording Tasman's discoveries, this map shows California as an island, which was not unusual in the late 17th century. The intricate cartouches and delicate colouring in this de Wit map are typical of the golden age of Dutch cartography.

Detail, right

This cartouche by the great engraver de Wit features Neptune in his shell chariot with a two-pronged 'trident'. Above Neptune is a portrait of Magellan.

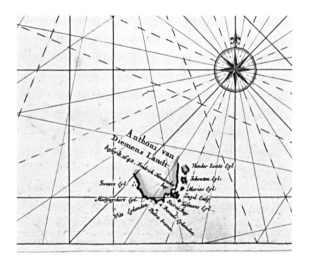

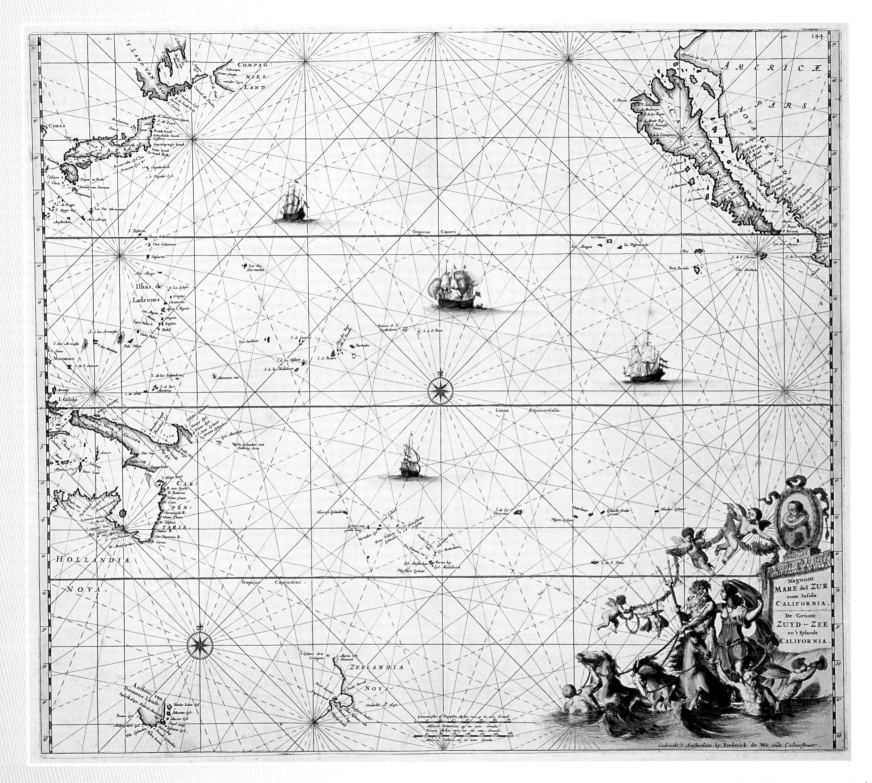

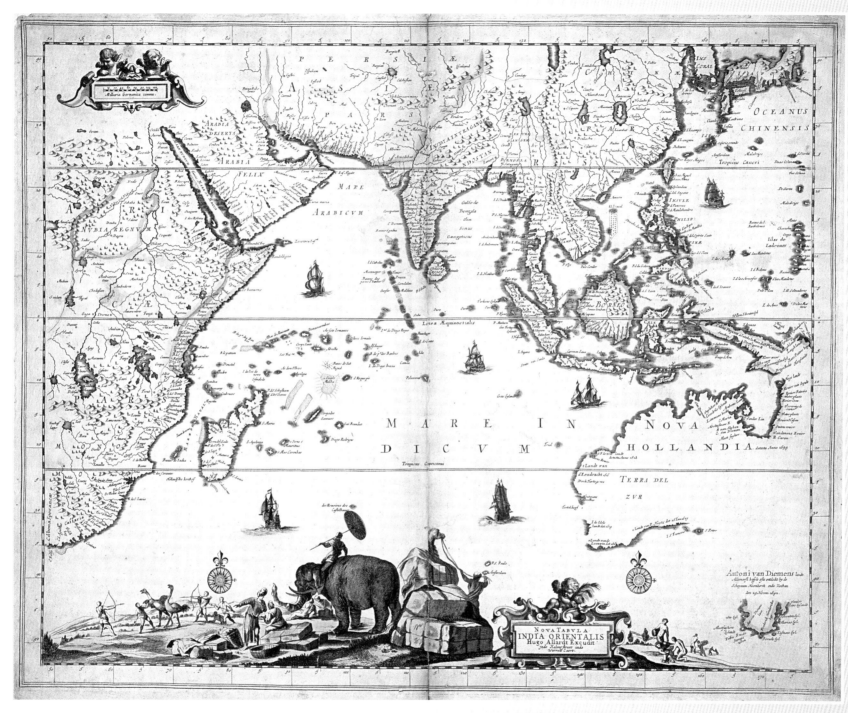

Nova Tabula India Orientalis

c. 1690 **Allardt**[6]

Apart from its elaborate decoration, this rare and attractive map contains considerable detail. Early explorers and dates are recorded, including Tasman's ships, *Heemskerck* and *Zeehaen*, shown as being in Van Diemen's Land on '29 Novem 1642'.

French Cartographers

First published in 1720 without any hypothetical coastlines, this reissue of de l'Isle's map adds dotted lines joining Van Diemen's Land, New Guinea and Australia. The large area depicted as 'Terre Australe du St. Esprit', (now Vanuatu), is the result of explorer, Pedro Fernandez de Quiros, claiming in 1606 that this was part of *Terra Australis*. This issue was published in Amsterdam by Covens & Mortier.

Detail

Before Cook's first voyage in 1768—1772, the east coast of Australia was a mystery. Some mapmakers believed hypothetical conjecture was preferrable to blank spaces.

By the end of the 1600s, Dutch interest in exploration had declined, and France replaced the Netherlands as the centre of cartography. Now scientific accuracy became more important than elaborate decoration. Guillaume de l'Isle was the leading French cartographer during this period of transition, which some claim to be the beginning of modern mapmaking.

HEMISPHERE ORIENTAL

Translation from the French text:

The Southern Hemisphere showing more clearly the Southern Lands by Guillaume de l'Isle of the Royal Academy of Sciences. [This map shows] *the new discoveries made in 1739 to the south of the Cape of Good Hope* [and was] drawn up by order of the management of the Indies Company, based on the reports and the original map of Monsieur de Lozier Bouvet who headed this expedition.

Extract from 'Voyage to the South Lands'

On the 19th July 1738, the two frigates L'Aigle and La Marie *leave the port of L'Orient. On the 8th September, they pass the Equator. On the 11th of October, arrive at the Isle of St. Catherine on the coast of Brasil. On the 13th November, set sail from this island to go in search of latitude 44° by 355° longitude. On the 26th,* [encounter] *thick fog at latitude 35° and* longitude 344°. Often it was impossible to distinguish objects beyond the firing range of a musket. The fog lasted until the 20th January.

As from the 3rd December big whales and strange birds could be seen from Gouémonde at 39° 20' latitude and 351° longitude. Thinking to be near land, sounded without finding the bottom at 180 fathoms. On the 7th experienced cold weather even though it was summer and the sun was nearing the solstice.

On the 10th, latitude 44° and to the south of Iron Island, the "Land of Sight" is placed in this spot by some geographers. The only land sighted was poorly positioned or an island. The 15° latitude and 7° longitude is the same as that of Paris! The air is very cold and saw 'the first icebergs which made us suspect that land is nearby. On the 21st [December], latitude 51° 23'. longitude 15° 22', variation noticed from 24° N.E. and 50° N.W. Compasses are giving different variations, an irregularity already experienced when approaching the icebergs in Hudson's Bay and Davis Strait.

L'Hemisphere Meridional
1740 de l'Isle

First published by de l'Isle in 1714, this map has been updated and reissued with new information including more recent explorers' routes. There is a block of French text on the left, and Dutch on the right. This issue was published in Amsterdam by Covens & Mortier.

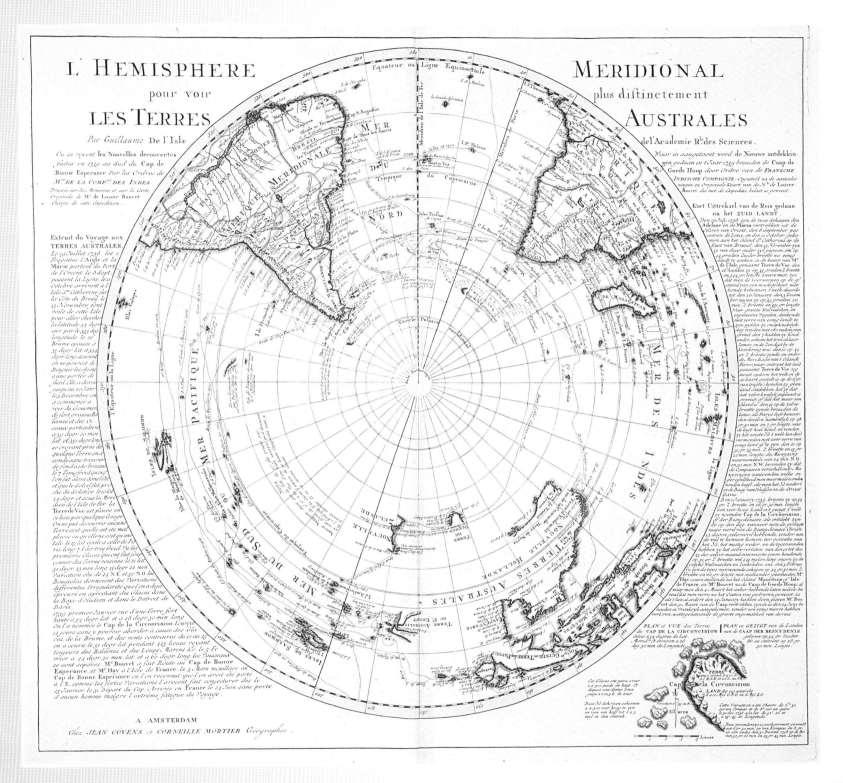

Orbis Vetus in utrâque continente

1752 **Robert de Vaugondy**[8]

This fascinating 18th century map records only ancient place names. Classical Latin texts are scattered throughout, suggesting an attempt to reconcile recent geographical discoveries with cultural heritage. The finely drawn and delicately coloured illustrations are a reminder of the earlier golden age of Dutch cartography.

Details

Two delicately coloured and finely drawn cartouches.

Text of lower detail:

Warning: Lest anyone be surprised, if he sees in this map of the whole world the island of Atlantis facing our America, we have considered that he should be advised that we have done this deliberately, since in the past not only the more celebrated geographers have restored this to antiquity but also Guillaume Sanson, the most learned of all, has penned that it was divided not inelegantly into ten kingdoms (adjoining the American regions) next to the ten sons of Neptune. Although all of these might seem established by no more than conjecture, nevertheless Chinese writings confirm by their claims that another continent was known to them, 400 years after the appearance of Christ; a scholar learned in oriental languages will publish these writings soon.

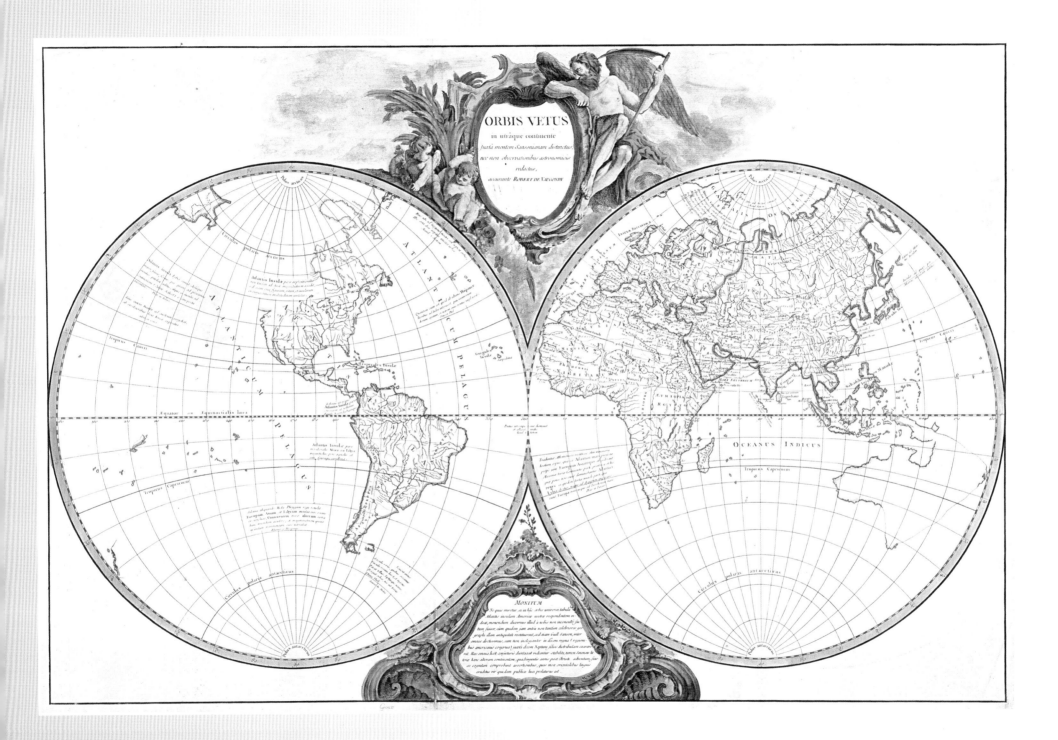

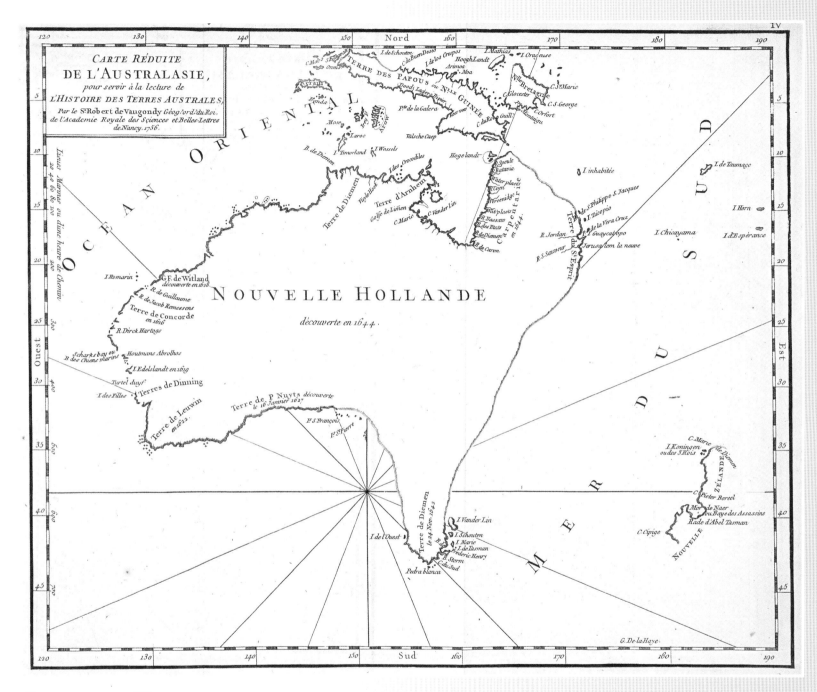

CARTE RÉDUITE
DE L'AUSTRALASIE,
pour servir à la lecture de
L'HISTOIRE DES TERRES AUSTRALES,
Par le Sr Robert de Vaugondy Géogr. ord. du Roi.
de l'Académie Royale des Sçiences et Belles-Lettres
de Nancy. 1756.

OCEAN ORIENTAL

NOUVELLE HOLLANDE

découverte en 1644.

MER DU SUD

NOUVELLE ZELANDE

Carte réduite de l'Australasie

1756 Robert de Vaugondy

Featuring 'Nouvelle Hollande' this map includes Tasman's charting of Van Diemen's Land and in northern Australia. 'Keerweer', named by Janszoon in 1606, is incorrectly located in New Guinea. Hypothetical coastlines join 'Terre de P. Nuyts' to Van Diemen's Land then to 'Terre du St. Esprit' (now Vanuatu) and across to the tip of 'Carpentarie' (Cape York). Despite contemporary uncertainty, there is a clearly defined, wide strait between Cape York and New Guinea.

Captain Cook carried this French map during his first voyage (1768–1771).

European Cartographers after Cook

With the information from Tasman's 1642–1643 and 1644 expeditions added to existing maps, most of the north, west and south-western coasts of Australia were charted, including the southern part of Van Diemen's Land. But with commercial exploitation replacing exploration, it was not until Captain James Cook's first voyage from 1768 to 1771 that the east coast of mainland Australia was finally defined.

Sailing west from New Zealand, Cook sighted the south-east coast of Australia on 19 April 1770. Naming the landfall Point Hicks, Cook proceeded north, comprehensively surveying the previously uncharted east coast. Reaching the tip of Cape York, he turned west, passing through, and thus confirming, the existence of Torres Strait.

Cook made two further voyages, (1772–1775 and 1776–1779), but it was his first expedition which had the greatest impact on the mapping of Australia.

Carte de l'Hemisphere Austral
1785 Benard/Cook[9]

An exact copy engraved by Robert Benard of Cook's 1777 map of the Southern Hemisphere. The French quickly translated and published accounts of Cook's voyages, reproducing the maps precisely. Published in *Abregé de l'Histoire Générale des Voyages*.

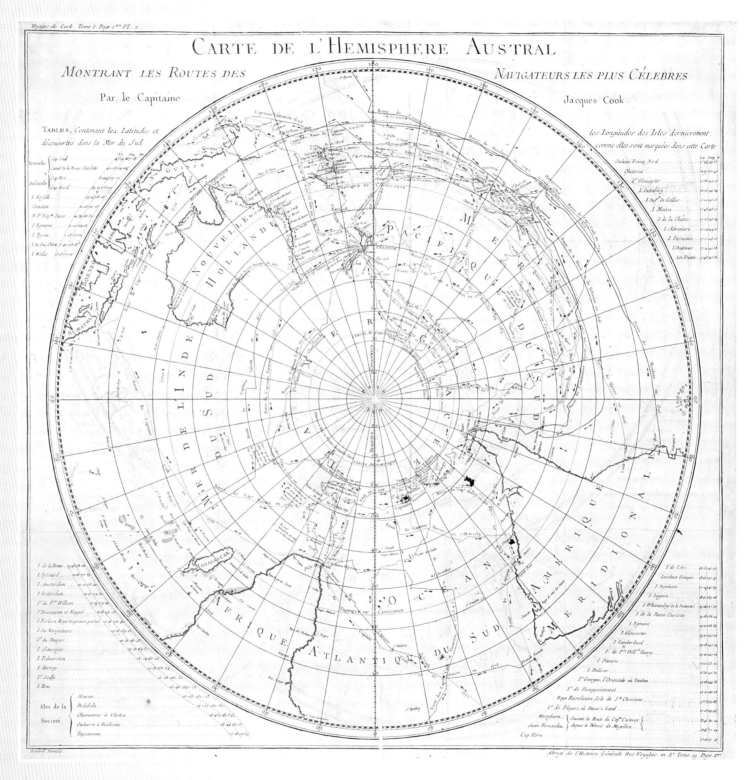

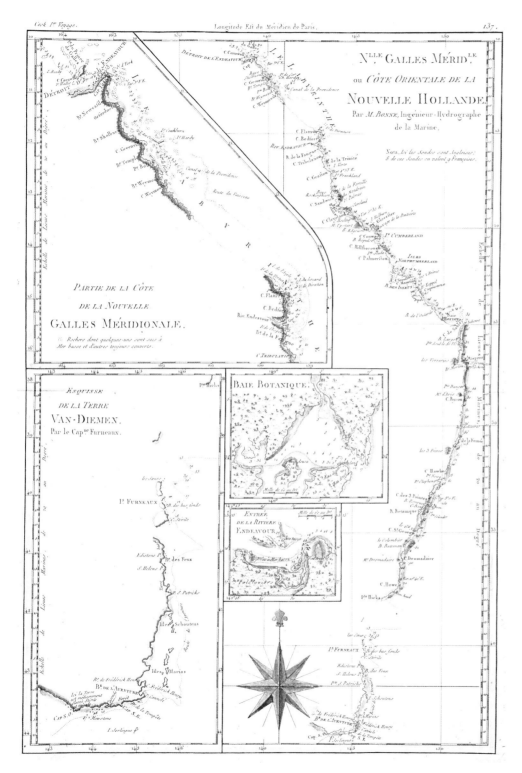

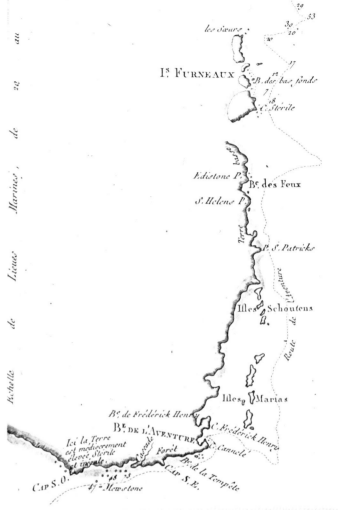

Nlle Galles Mérid'le ou Côte Orientale de la Nouvelle Hollande

1788 Bonne[10]

A French copy of charts from Cook's voyages. Three of the insets — Cape York, Botany Bay and Endeavour River — are from his first expedition. The Van Diemen's Land inset is from his second voyage. Published in volume II of Bonne's *Atlas Encyclopédique*, 1788.

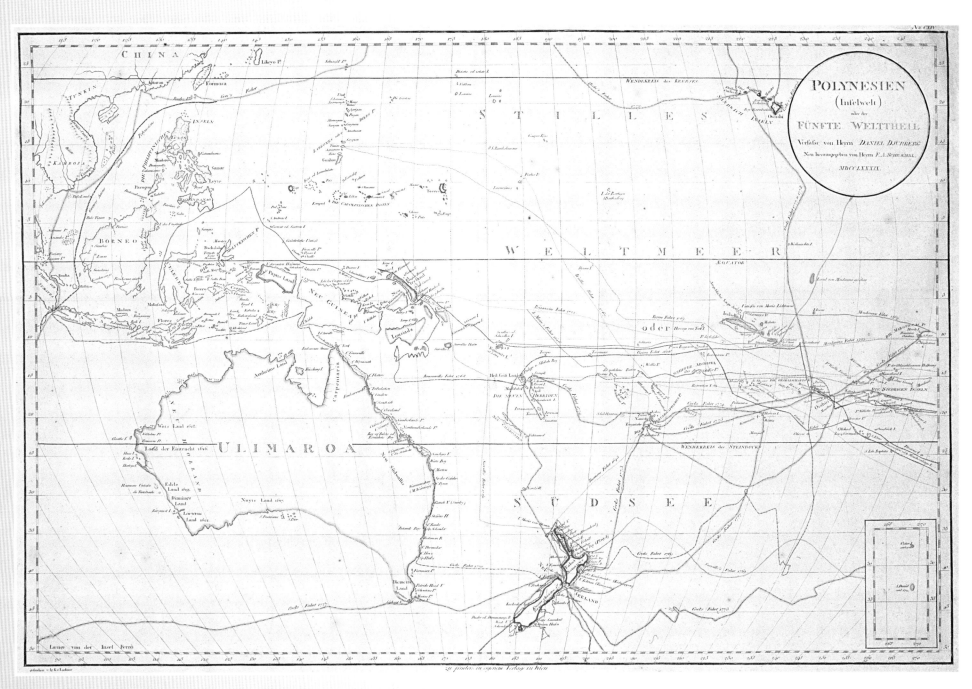

Polynesien

1789 Djurberg[11]

During Cook's first voyage, the Maoris told him that they believed there was a land to the north-west, called 'Ulimaroa'. Djurberg was the first mapmaker to adopt this Maori word to describe Australia.

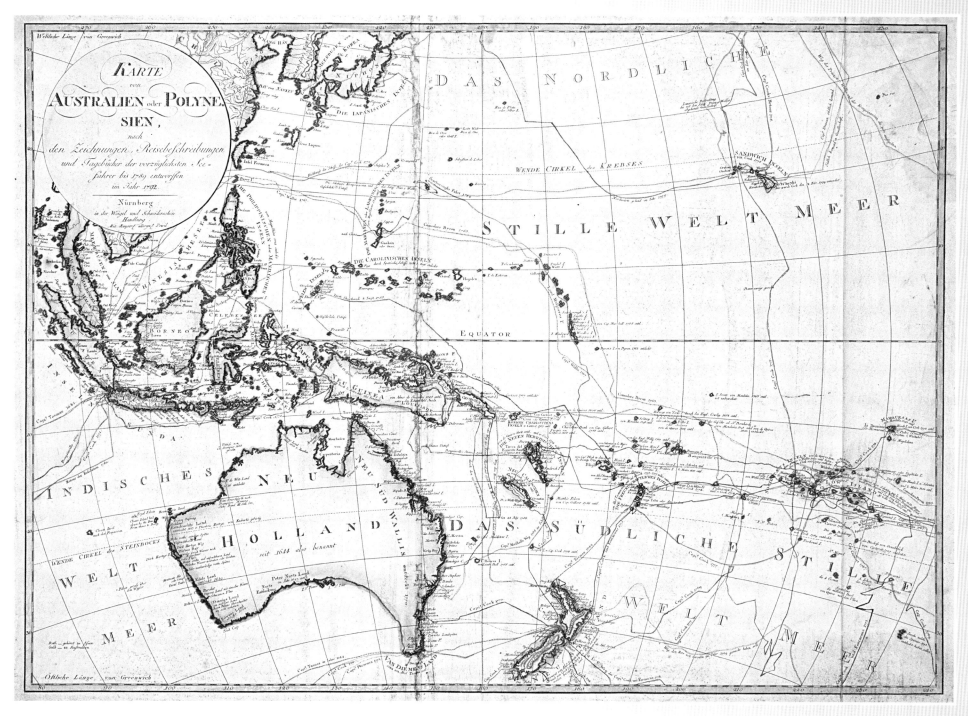

Karte von Australien oder Polynesien

c. 1792 **Weigel und Schneiderschen** [SIC][12]

This German map includes Tasman's, Furneaux's and Cook's tracks south of Tasmania.

La Nuova Olanda e La Nuova Guinea

1798 Cassini[13]

This map features Cook's 1770 charting of the east coast of Australia. The designated position of the reef 'los Trials' well off the north-west coast of Western Australia is believed to be the result of deliberate misinformation. In 1622 the British ship *Trial* was wrecked with the loss of more than 90 lives. To minimise his negligence, the captain, James Brookes, falsified his records when indicating the wreck's location. *Trial* was finally correctly located in 1969 off the Monte Bello Islands, many miles to the east of Brooke's reported position. The *Trial* is the earliest recorded shipwreck in Australian waters. Note that New Guinea is named 'Land of Parrots'.

La Nuova Olanda e La Nuova Guinea

1798 **Cassini**

Details

Van Diemen's Land: Defining the Island

After Tasman's 1642 visit, when he charted part of the island, 130 years would pass before cartographers were able to add to this scant outline. In 1772, James Cook left England on his second voyage (1772–1775). The captain of the expedition's second ship, *Adventure*, was Tobias Furneaux. On 8 February 1773, having lost contact during fog in the Antarctic, both ships sailed separately for New Zealand, with Furneaux deciding to explore Van Diemen's Land on the way. He spent some days in a bay he named 'Adventure' after his ship, then proceeded up the east coast past 'Furneaux's Isles' and into Bass Strait. However, having land in sight for most of the way, he believed Bass Strait was probably a large bay. Furneaux then turned east for New Zealand to rejoin Cook. Cook himself did not visit Van Diemen's Land during this voyage.

During Cook's third voyage (1776–1779) in the *Resolution*, he made his first and only visit to Van Diemen's Land. On 26 January 1777, Cook anchored in Adventure Bay, which had been visited by Furneaux four years earlier. After four days spent collecting wood, water and fodder for the animals, he proceeded east to New Zealand. Cook did not return to England from this voyage. He was killed in Hawaii on 14 February 1779.

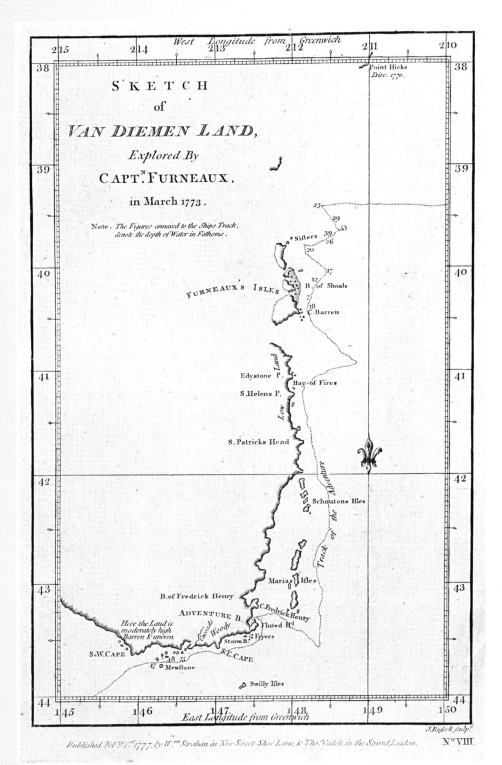

Sketch of Van Diemen Land

1777 Furneaux

This is believed to be the first printed English map to depict separately part of the coast of Van Diemen's Land. The track of the *Adventure* is shown approaching South West Cape then tracking up the east coast. 'Point Hicks Disc. 1770' in the top right corner refers to Cook's New Holland landfall in 1770.

Made by Captain Furneaux in 1773 and published in 1777 by William Strahan as chart no. VIII in *A Voyage Towards the South Pole and Round the World*, the narrative of Cook's second voyage.

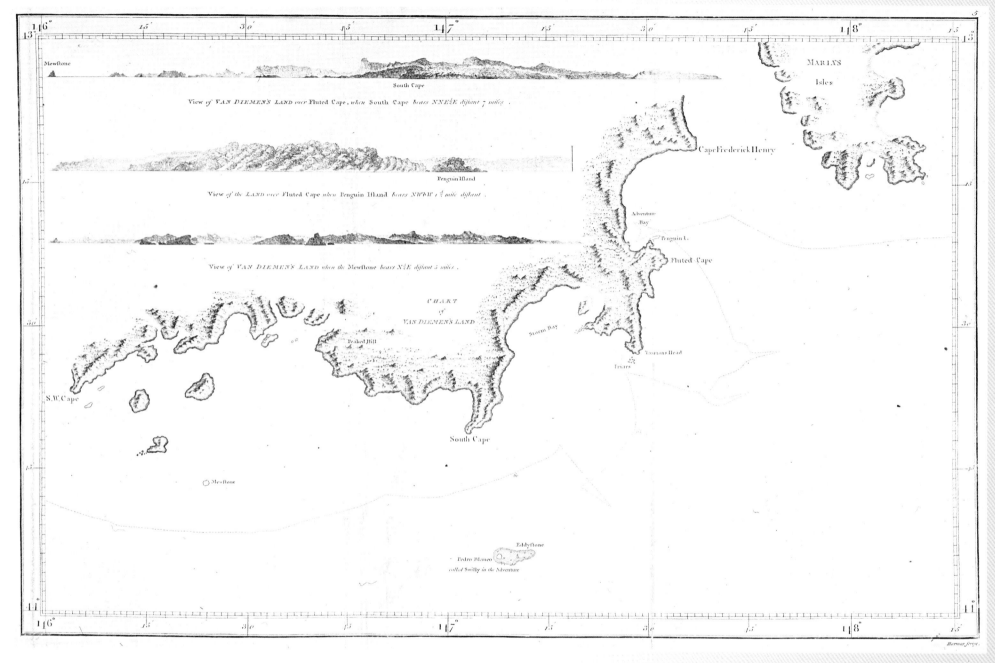

Chart of Van Diemen's Land

1785 **Cook**

Map of the south-east coast made in 1777 during Cook's first and only visit to Van Diemen's Land. The track of the *Resolution* is shown rounding South West Cape then proceeding to anchor in Adventure Bay. Unaware that he was actually on an island (Bruny), Adventure Bay is depicted as part of mainland Van Diemen's Land, with Tasman Peninsula identified as 'Maria's Isles'. Published in *A Voyage to the Pacific Ocean*, 1785, the official narrative of Cook's third voyage.

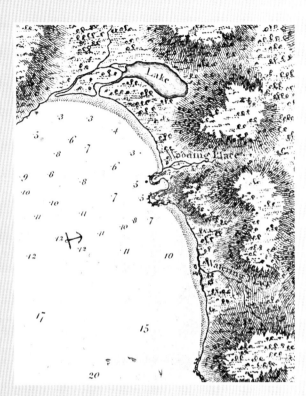

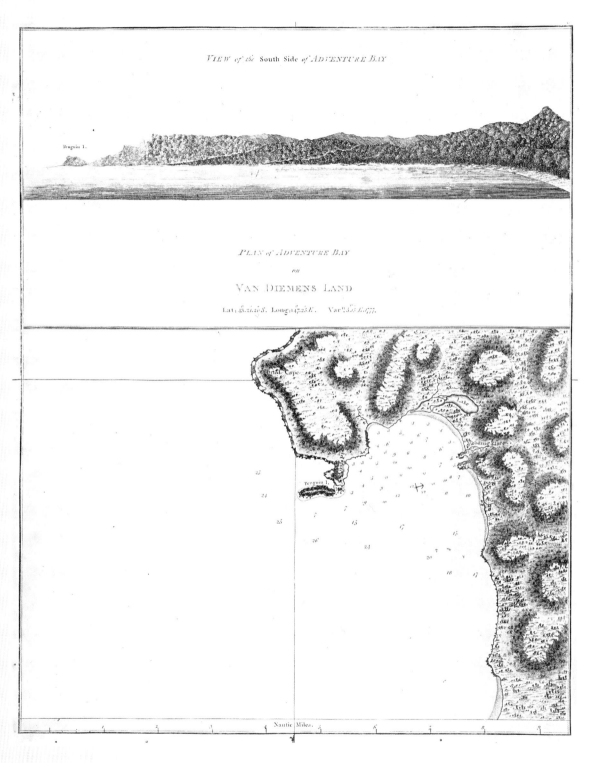

Plan of Adventure Bay on Van Diemens Land

1785 **Cook**

A large, finely drawn silhouette of Adventure Bay, below which the map indicates Cook's anchorage, and soundings. Also noted are 'wooding' and 'watering' places. Orientated with south at the top. Published in *A Voyage to the Pacific Ocean*, 1785, the official narrative of Cook's third voyage.

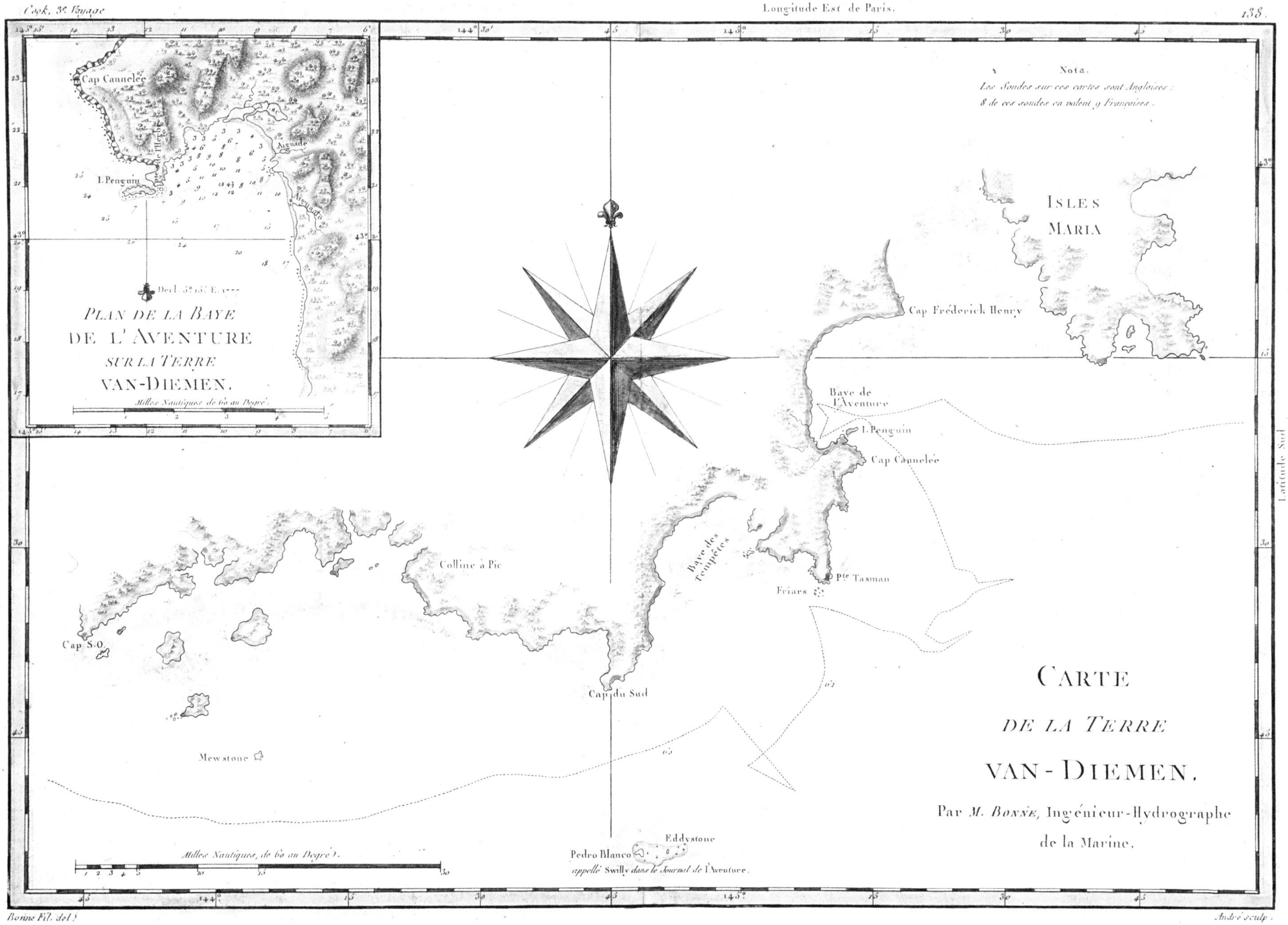

Nota.

Les Sondes sur ces cartes sont Angloises :
8 de ces sondes en valent 9 Françoises.

ISLES MARIA

Cap Frederick Henry

Baye de L'Aventure

I. Penguin

Cap Cannelée

Colline à Pic

Baye des Tempêtes

Pte Tasman

Friars

Cap S.O.

Cap du Sud

Mewstone

Latitude Sud

CARTE

DE LA TERRE

VAN-DIEMEN.

Par M. BONNE, Ingénieur-Hydrographe

de la Marine.

Eddystone

Pedro Blanco
appellé Swilly dans le Journal de l'Aventure.

Milles Nautiques, de 60 au Degré.

Inset:

Cap Cannelée

I. Penguin

Aiguade

Decl. 5° 15' E.

PLAN DE LA BAYE
DE L'AVENTURE
SUR LA TERRE
VAN-DIEMEN.

Milles Nautiques de 60 au Degré.

Bonne Fil. del.   André sculp.

Carte de la Terre Van-Diemen

1788 **Bonne/Cook**

This is a French copy of Cook's maps. The inset of Adventure Bay is orientated with south at the top.

Published in Bonne and Desmarest's *Atlas Encyclopédique*, 1788.

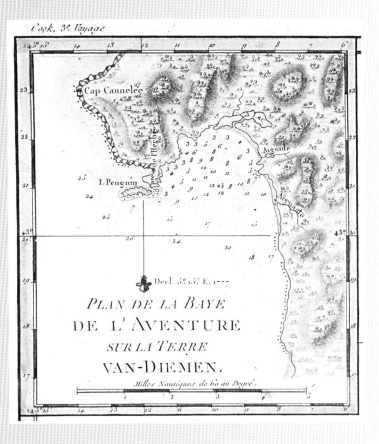

After the establishment of the penal settlement at Sydney Cove in 1788, the British began exploring south towards Van Diemen's Land. By 1798, Dr George Bass had reached Westernport in an open whaleboat and upon his return to Sydney, reported that ocean swells indicated the existence of a strait between New Holland and Van Diemen's Land. This was confirmed in 1798–1799 when Matthew Flinders, together with Bass, surveyed the strait, and successfully circum-navigated Van Diemen's Land, recording, for the first time, the outline of the island. Apart from proving Bass

Strait existed, this significant maritime expedition also opened a shorter shipping route between England and Port Jackson (Sydney Harbour).

Meanwhile, in April 1792, Bruni d'Entrecasteaux in the ships *Recherche* and *Espérance* had anchored in and named Recherche Bay, returning in January 1793 for a second visit. Shortly after, in April 1793, Captain John Hayes, on a private voyage from India to New Guinea, was forced by adverse winds towards Van Diemen's Land, where he decided to enter and explore what he believed was an unmapped river.

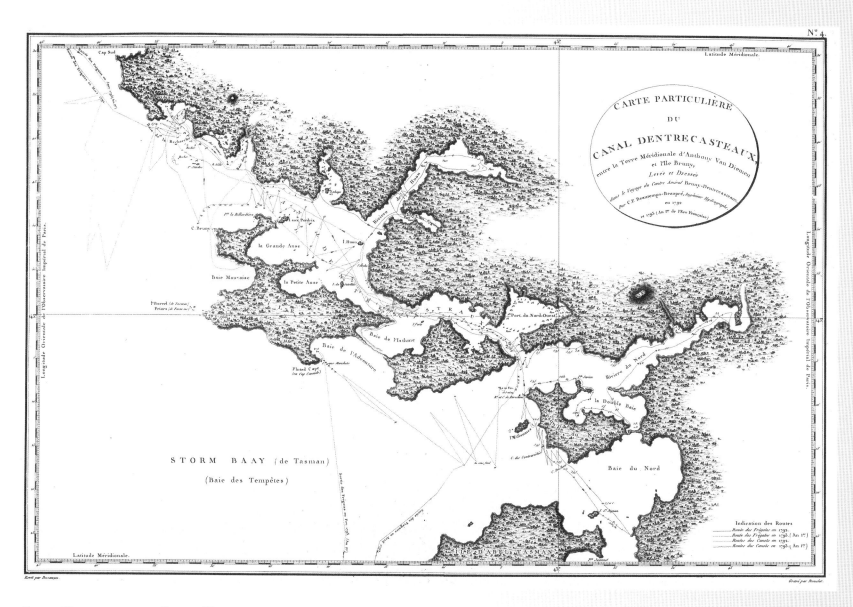

Carte Particulière du Canal Dentrecasteaux

1807 **Beautemps-Beaupré**[14]

In 1792–1793, Bruni d'Entrecasteaux's hydrographer, Beautemps-Beaupré, carried out detailed exploration and mapping of the D'Entrecasteaux Channel, the Derwent estuary and its environs. Recherche Bay, Huon River, Port Esperance and D'Entrecasteaux Channel are among surviving names. The River Derwent, however, is designated 'Rivière du Nord'. Published in *Atlas du Voyage de Bruny-Dentrecasteaux*, 1807.

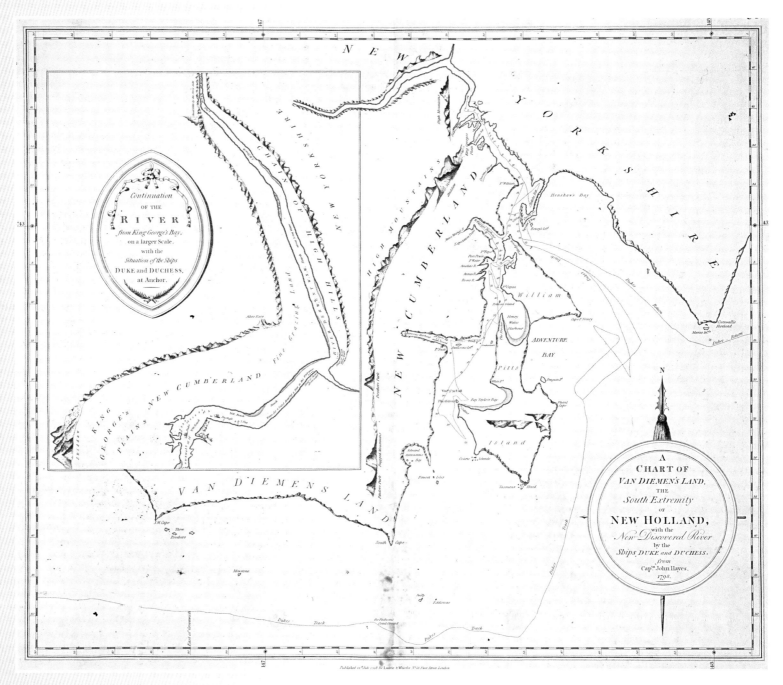

A Chart of Van Diemen's Land, the
South Extremity of New Holland

1798 Hayes[15]

In April 1793, with his two ships, *Duke* and *Duchess*, Captain John Hayes entered and named the River Derwent, after the river in his hometown in Cumberland. Believing he was the first European to explore the river and its environs, he later discovered that d'Entrecasteaux had sailed into the Derwent estuary only a few months before him.

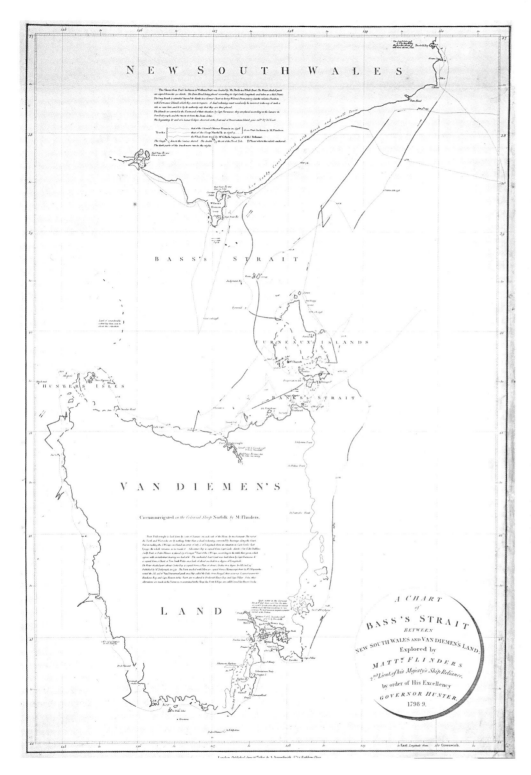

A Chart of Bass's Strait between New South Wales and Van Diemen's Land

1801 A. Arrowsmith[16]

This historic map shows for the first time the full outline of Van Diemen's Land and records Flinders's survey made during his and Bass's circumnavigation of the island. Departing Port Jackson in the *Norfolk* on 7 October 1798, they returned on 11 January 1799, having charted the previously unexplored areas of the coastline and proving the existence of Bass Strait. Flinders acknowledges the contribution of earlier explorers such as Furneaux, Cook and Hayes.

Although dated 1800, this is a second revised edition published in Flinders's book *Observations on the Coast of Van Diemen's Land*, 1801.

Detail

The Tamar River.

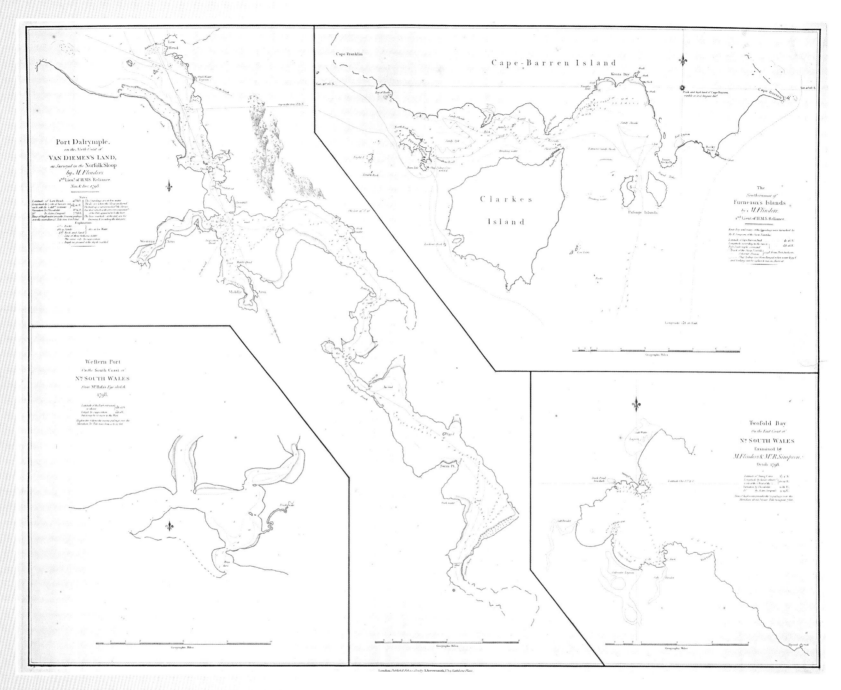

Port Dalrymple, Furneaux's Islands, Twofold Bay and Western Port

1801 A. Arrowsmith

These maps include surveys undertaken by Flinders during the *Norfolk*'s circumnavigation of Van Diemen's Land in 1798–1799 and a sketch of the Tamar River as far as Swan Point. George Bass made the 'eye sketch' of Western Port during his whaleboat voyage earlier in 1798, noting that despite his best endeavours, the map was far from perfect.

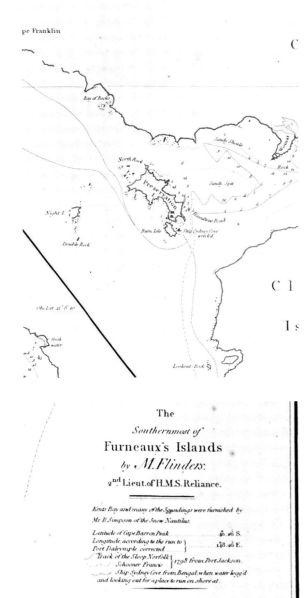

The
Southernmost of
Furneaux's Islands
by M. Flinders.
2nd Lieut. of H.M.S. Reliance.

Kents Bay and many of the Soundings were furnished by
Mr R. Simpson of the Snow Nautilus.

Latitude of Cape Barren Peak .... 40. 26 S.
Longitude, according to the run to ⎫ 148.26 E.
Port Dalrymple, corrected .... ⎭
Track of the Sloop Norfolk ⎫ 1798 from Port Jackson.
Schooner Francis ⎭
Ship Sydney Cove from Bengal, when water logg'd
and looking out for a place to run on shore at.

Port Dalrymple, Furneaux's Islands,
Twofold Bay and Western Port
1801 **A. Arrowsmith**

Details

In February 1797, en route from India to Port Jackson and in imminent danger of sinking, the *Sydney Cove*, under the command of Captain Hamilton, was beached on a little island which he named Preservation Island. A longboat with 17 men set out for help, but was wrecked on the Victorian coast. Only three men survived the trek of more than 600 km to Sydney where a rescue was mounted, and two vessels, *Eliza* and *Francis*, were despatched to the island. *Francis*, with 25 survivors, returned safely to Sydney, but *Eliza*, with 12 survivors aboard, was never heard from again.

'Track of the Sloop Norfolk' refers to Bass and Flinders's 1798–1799 circumnavigation of Tasmania.
Track of the 'Schooner Francis' refers to *Francis*'s third salvage voyage to Preservation Island in January 1798.

Louis de Freycinet's Maps

Louis de Freycinet, a French naval officer, was the cartographer-geographer in Nicolas Baudin's expedition which left Le Havre in 1800 in the ships *Le Géographe* and *Le Naturaliste* to survey the uncharted coasts of southern Australia. To pre-empt any possibility of French territorial claims, the British ordered Matthew Flinders to carry out the same task. After Baudin's visit to Sydney, in 1803 a small settlement was established at Risdon Cove to demonstrate British sovereignty over the Bass Strait region and Van Diemen's Land. Later, becoming aware that the French had named a long section of the South Australian and Victorian coastlines 'Terre Napoléon', the British had their suspicions further aroused.

From this expedition, Freycinet made a series of maps including some of Van Diemen's Land. Freycinet returned to Paris in March 1804 where he later became involved in their publication. Baudin had died on Mauritius in September 1803. *La Casuarina*, referred to in the maps, was a schooner acquired in Sydney by Baudin for inshore surveying.

These maps by Freycinet appeared in *Voyage de Découvertes aux Terres Australes*, tome 3, Paris, 1812.

Carte Générale du Détroit de Bass
1812 Freycinet

Bass Strait was important strategically to the British, reducing the sailing time to Sydney from England. The known anchorages on either side of the strait are shown: 'Port Western', east of present day Port Phillip, and Port Dalrymple on the Tamar River. The tracks of Baudin's ships are shown and his name for the coast 'Terre Napoléon' is in the top left corner.

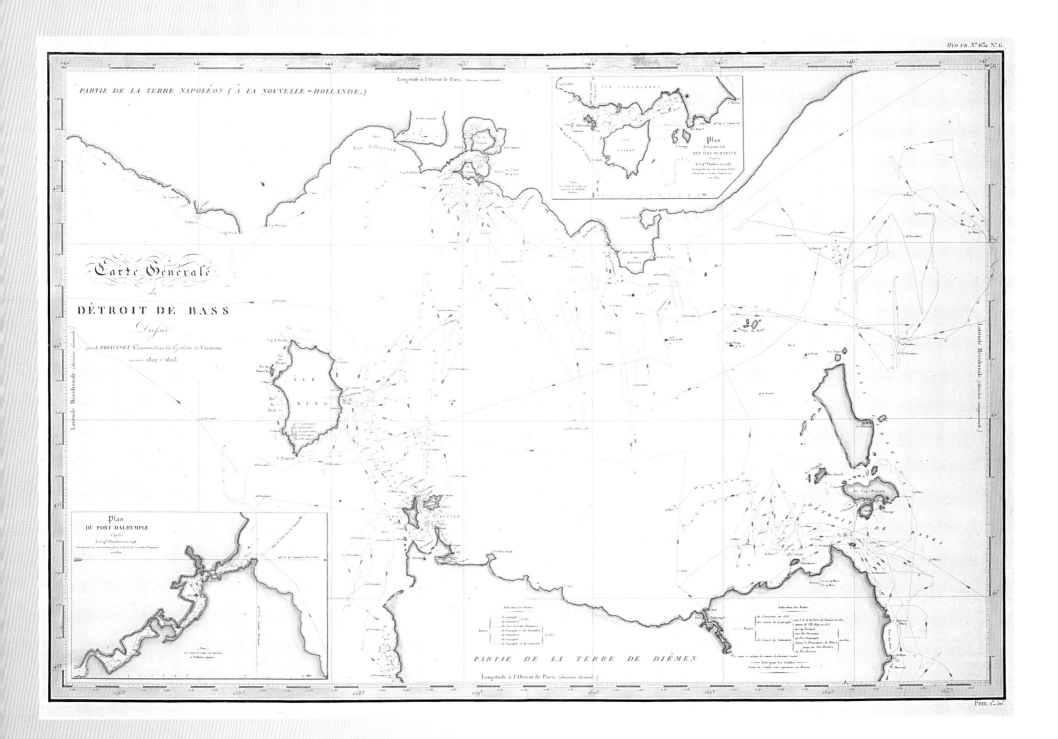

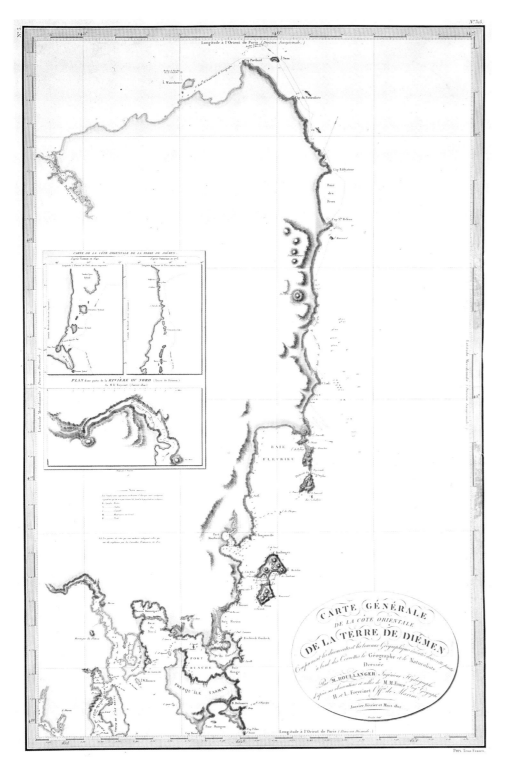

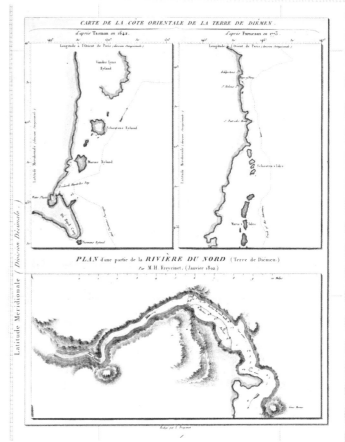

Carte Générale de la côte orientale de la Terre de Diémen

1812 **Freycinet**

Map of the coastline of Van Diemen's Land from North Bruny Island to the Tamar River.

Detail

The top two insets acknowledge earlier maps by Tasman (1642), and Furneaux (1773). The third inset is an enlargement of the River Derwent, north of Risdon Cove. This was drawn by Louis' brother, Henri, also a member of Baudin's expedition.

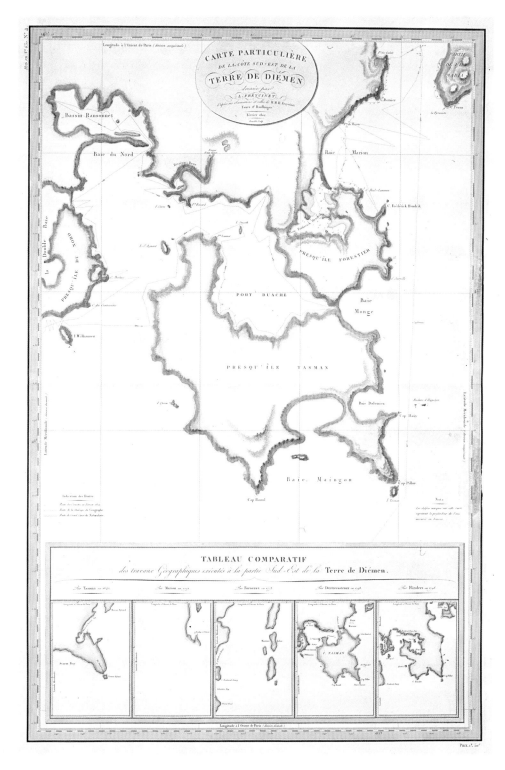

Carte Particulière de la côte sud-est
de la Terre de Diémen

1812 **Freycinet**

In this map of part of south-east Van Diemen's Land,
Forestier Peninsula has rather a squashed appearance.
Five insets of previous maps of the area are attributed to
appropriate earlier explorers.

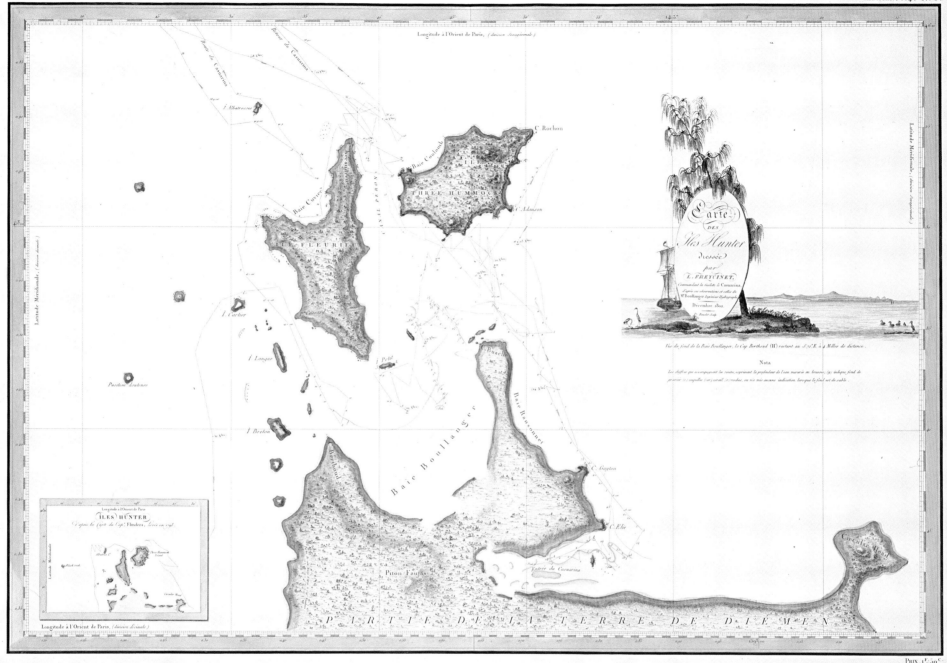

Carte des Iles Hunter

1812 **Freycinet**

Freycinet's map of the islands off the north-west tip of Van Diemen's Land includes Hunter (outlined in yellow) and Three Hummock Islands. Robbins Island is shown as part of the mainland of Van Diemen's Land, although parts of the coastline are left blank, suggesting Freycinet believed there was a bay or a passage between the mainland and Robbins Island.

Post-Settlement Australia and Oceania

By 1804, Van Diemen's Land is depicted internationally on maps as being separate from mainland Australia. French interest in the Pacific region is shown in Lapie's 1809 map of New Holland and Levasseur's 1838 decorative map of the Pacific Basin. The following translation of the French text from Levasseur's map *Océanie* (p. 52) provides an insight into how some Europeans viewed this region in the early 19th century.

Oceania, the fifth major region of the world, spans an immense area of ocean and a huge number of islands, encompassing more than half the globe. Australia proper, the fifth continent, was named New Holland by the Dutch when they arrived but is now called Australia and is vaster than Europe. Dumont-Durville divided Oceania into four large sections, 1st Malaysia, 2nd Melanesia or Australia, 3rd Polynesia and 4th Micronesia. The climate of Oceania is generally temperate except near the equator where it is consequently very hot. The mountains of the islands have many volcanoes; no other part of the world has more. There are numerous mines of gold, silver, lead, copper etc. Throughout the region nature has furnished a superabundance of edible vegetation of the Palm family, suitable for domestic use. Extremely plentiful, they are diffused through even the most remote and smallest islands. The inhabitants are copper and black and profess religions of idolatry and Mohammedanism. Catholicism, Calvinism and Christianity are followed by island colonies dominated by Europeans. Governments are absolute and barbaric. Inhabitants of some islands continue to practise cannibalism. Some however have a sweet character in singular contrast to the cruel ideas of other Oceanians. The total population of Oceania is 27,000,000.

General-Charte von Australien

1804 Reinecke[17]

One of the earlier maps to show Van Diemen's Land as separate from mainland Australia, it correctly delineates Bass Strait, and King, Hunter and the Furneaux Islands. This German map measures longitude west from Greenwith at the top of the map and the more common east from Greenwich at the bottom.

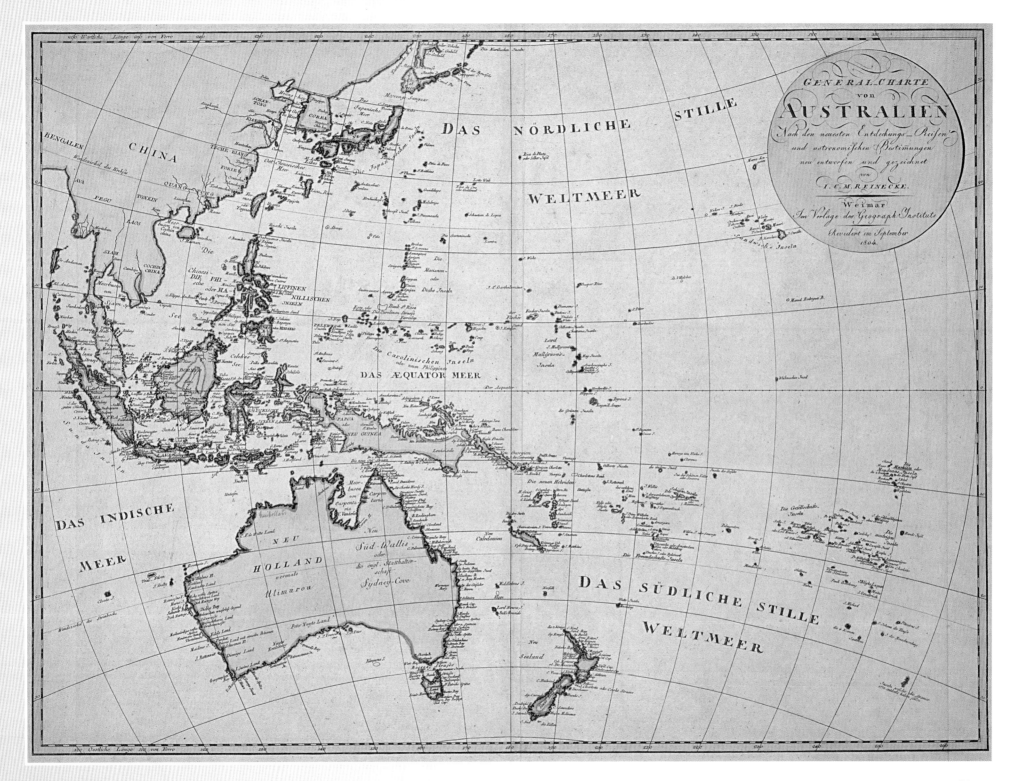

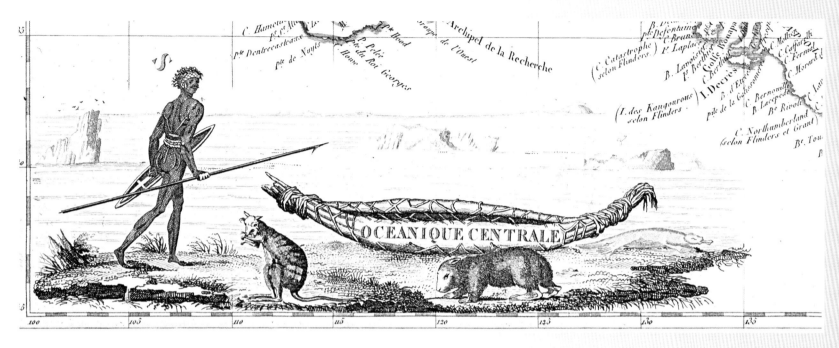

Océanique Centrale
1809 Lapie[18]

Lapie's map has a wealth of coastal placenames given by French explorers including 'Terre Napoléon'. An early complete map of the continent of Australia.

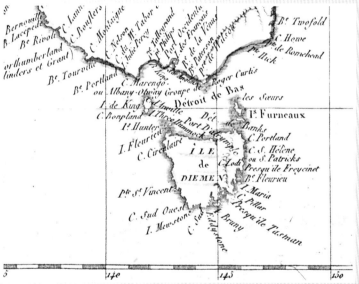

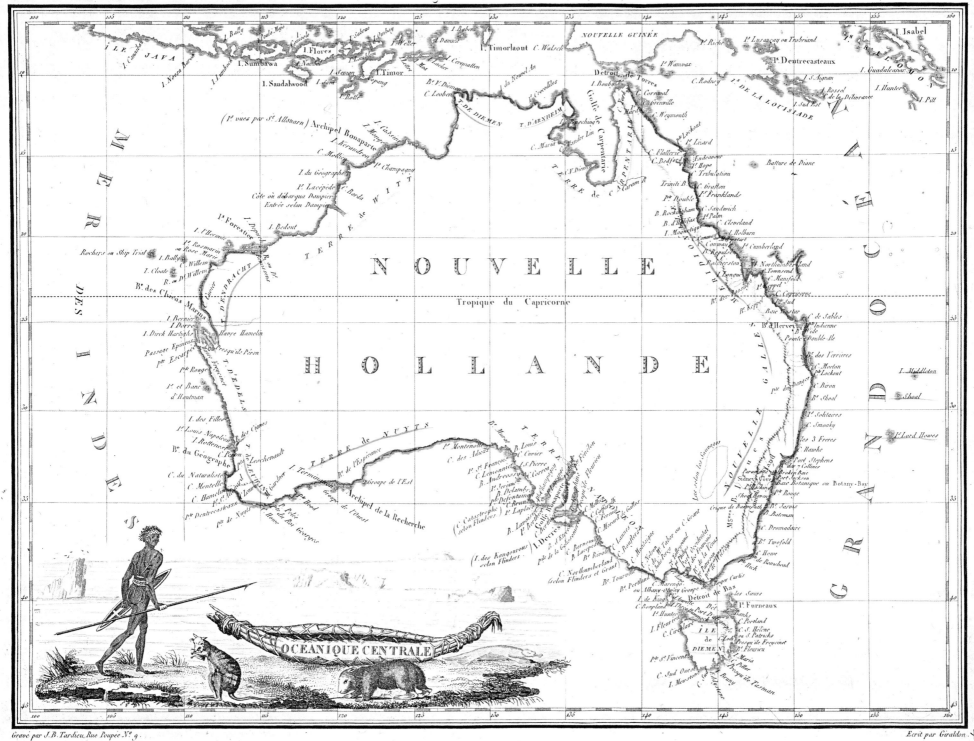

Gravé par J.B. Tardieu, Rue Poupée N.º 9.

Ecrit par Giraldon.

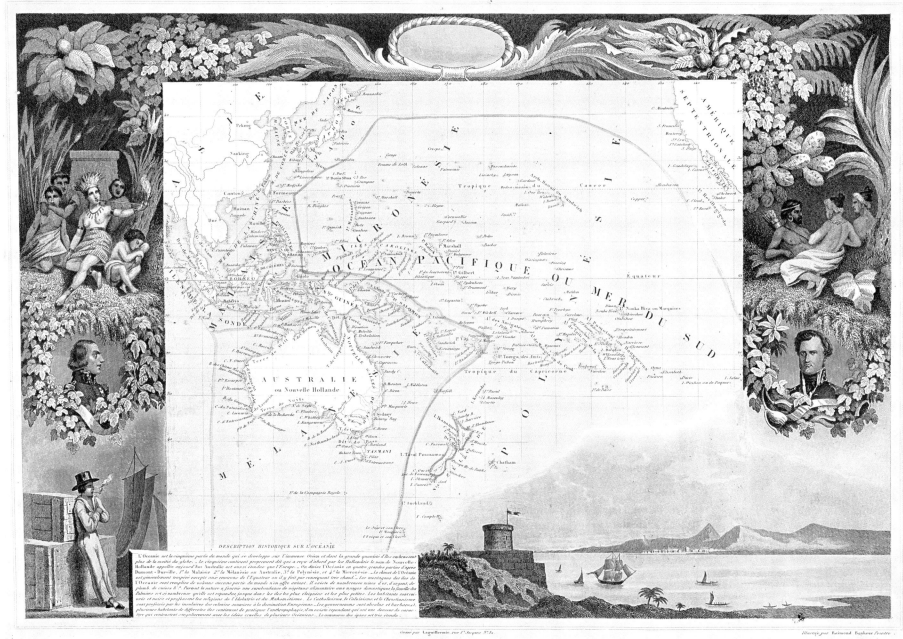

Océanie

1838 Levasseur[19]

With beautiful engravings, the islands of the South Pacific
are well illustrated, although the continent of Australia
appears relatively 'empty'.

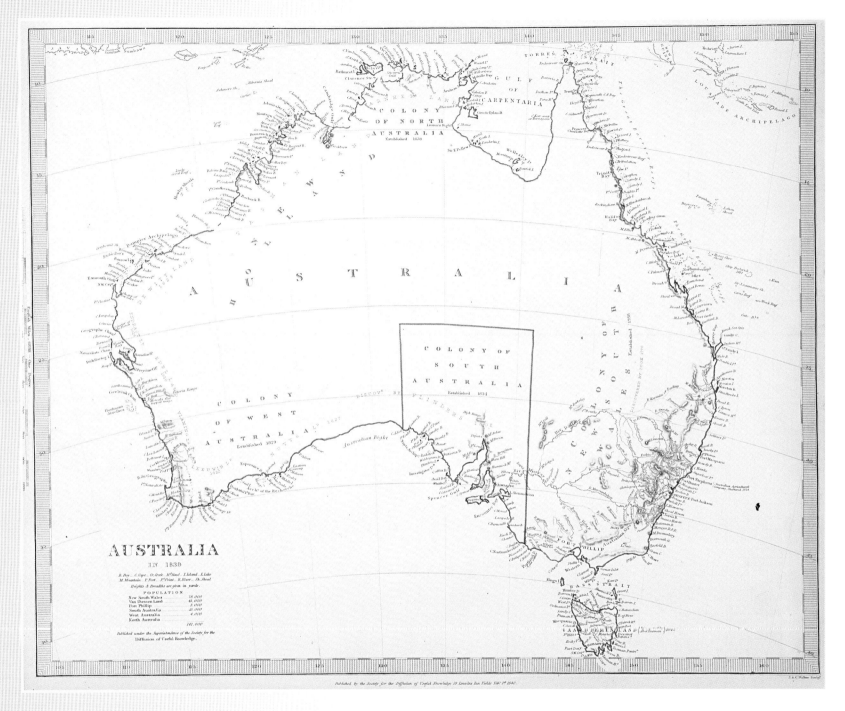

AUSTRALIA
IN 1839

B. Bay   C. Cape   Gr. Grit   Hd. Head   I. Island   L. Lake
M. Mountain   P. Port   Pt. Point   R. River   Sh. Shoal
Heights & Breadths are given in yards.

POPULATION
New South Wales 78,000
Van Diemen Land 43,000
Port Phillip 3,000
South Australia 13,000
West Australia 4,000
North Australia .... —
                          141,000

Published under the Superintendence of the Society for the
Diffusion of Useful Knowledge.

Australia in 1839

**1840 Society for the Diffusion
of Useful Knowledge**[20]

It is interesting that a 'Colony' of North Australia is shown
as being established in 1838, although it is listed as having

no settlers. A military outpost, established in 1838 at Port
Essington in northern Australia, probably caused the confusion.

Australia

LEFT-HAND SHEET OF A TWO-SHEET MAP OF AUSTRALIA

1848 J. Arrowsmith[21]

In this map, South Australia and Western Australia do not share a common boundary. The area between the borders was, presumably, part of New South Wales.

Eastern Portion of Australia

RIGHT-HAND SHEET OF A TWO-SHEET MAP OF AUSTRALIA

1848 J. Arrowsmith

Although 46 counties in New South Wales are tabulated, only 19 appear on the map (within the green border). This reflects Governor Darling's futile 1829 proclamation that settlement be limited to these 19 counties. Coloured outlines within Van Diemen's Land indicate 'Police Districts', established to assist in controlling both Aborigines and convicts. The first recorded European contact with Australia is noted west of Cape York.

Detail

Cape Keer-weer (or Turnagain), named by Captain Janszoon in 1606, is the earliest European placename in Australia in current use.

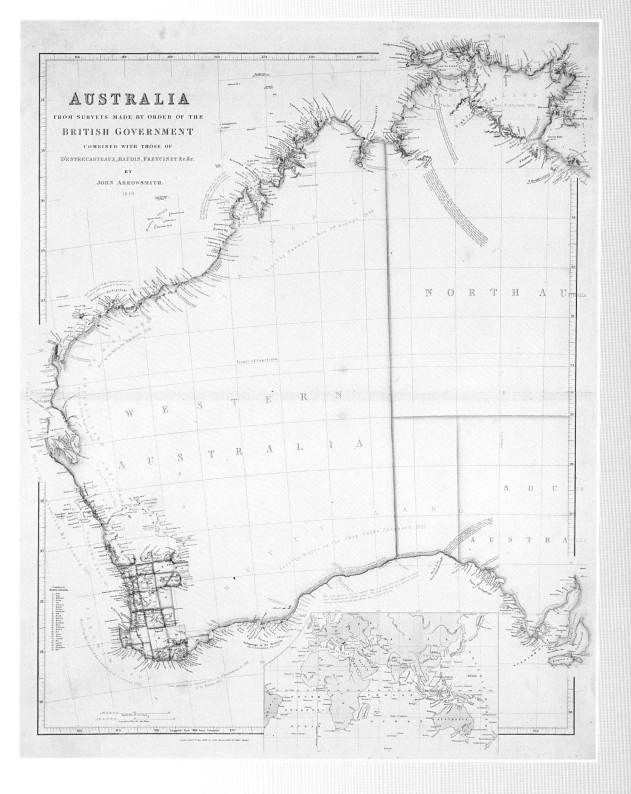

AUSTRALIA

FROM SURVEYS MADE BY ORDER OF THE

BRITISH GOVERNMENT

COMBINED WITH THOSE OF

D'ENTRECASTEAUX, BAUDIN, FREYCINET &c.&c.

BY

JOHN ARROWSMITH.

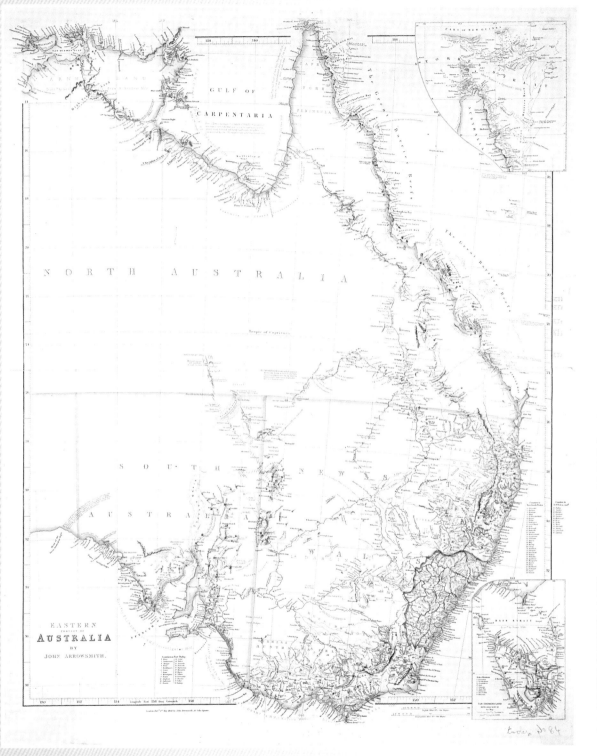

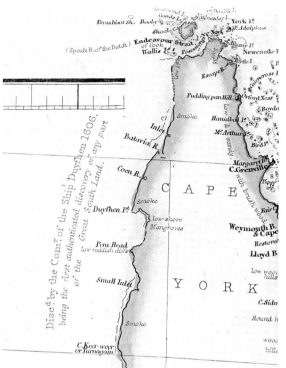

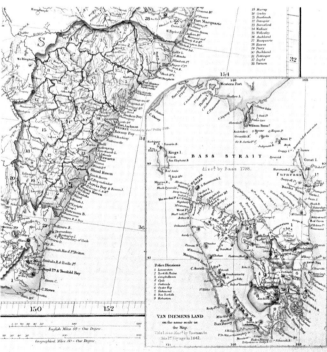

Australia

FIRST ISSUED **1851 Tallis**[22]

Originally published in Tallis's 1851 *Illustrated Atlas*, this is a later edition. It shows Victoria separated from New South Wales, with the new colony named 'Victoria or Port Phillip'. 'Australia Felix' was the name explorer Thomas Mitchell gave to the rich grazing lands south of the Murray River in what is now western Victoria.

NATIVES OF AUSTRALIA FELIX.

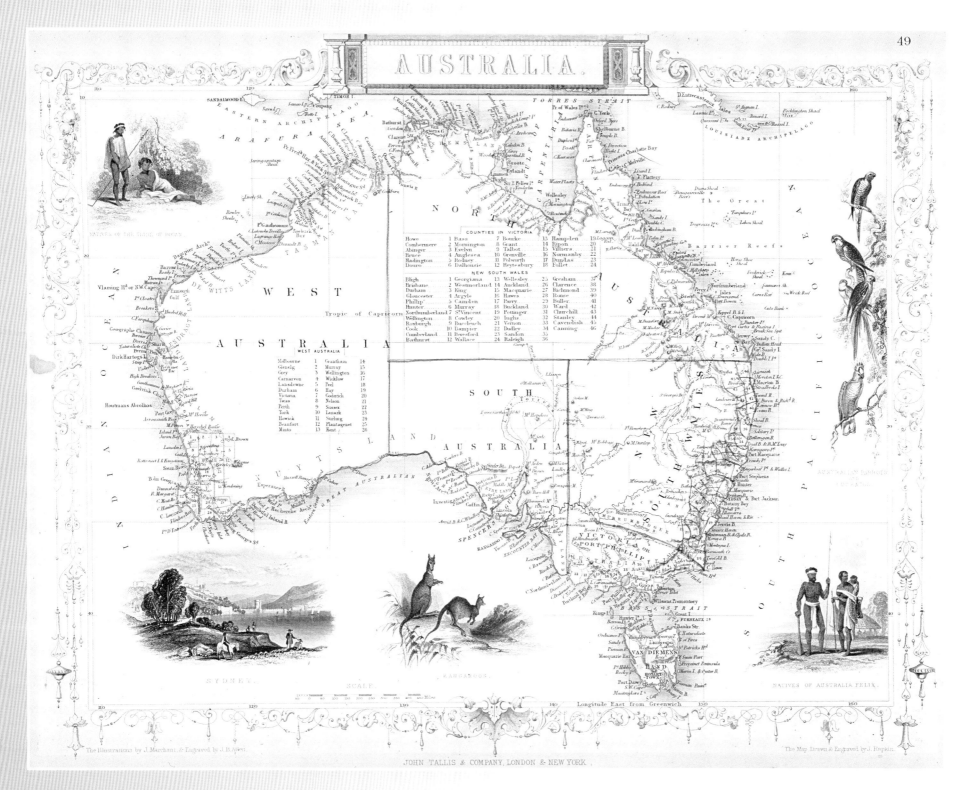

AUSTRALIA.

COUNTIES IN VICTORIA

Howe	1	Bass	7	Bourke	13	Hampden	19
Combermere	2	Mornington	8	Grant	14	Ripon	20
Abinger	3	Evelyn	9	Talbot	15	Villiers	21
Bruce	4	Anglesea	10	Grenville	16	Normanby	22
Hadington	5	Rodney	11	Polwarth	17	Dundas	23
Douro	6	Dalhousie	12	Heytesbury	18	Follet	24

NEW SOUTH WALES

Bligh	1	Georgiana	13	Wellesley	25	Gresham	37
Brisbane	2	Westmorland	14	Auckland	26	Clarence	38
Durham	3	King	15	Macquarie	27	Richmond	39
Gloucester	4	Argyle	16	Hawes	28	Rouse	40
Phillip	5	Camden	17	Parry	29	Buller	41
Hunter	6	Murray	18	Buckland	30	Ward	42
Northumberland	7	S.Vincent	19	Pottinger	31	Churchill	43
Wellington	8	Cowley	20	Inglis	32	Stanley	44
Roxburgh	9	Bucclench	21	Vernon	33	Cavendish	45
Cook	10	Dampier	22	Dudley	34	Canning	46
Cumberland	11	Beresford	23	Sandon	35		
Bathurst	12	Wallace	24	Raleigh	36		

WEST AUSTRALIA

Melbourne	1	Grantham	14
Genelg	2	Murray	15
Grey	3	Wellington	16
Carnarvon	4	Wicklow	17
Lansdowne	5	Peel	18
Durham	6	Bay	19
Victoria	7	Goolrich	20
Twiss	8	Nelson	21
Perth	9	Sussex	22
York	10	Lanark	23
Howick	11	Stirling	24
Beaufort	12	Plantagenet	25
Minto	13	Kent	26

SYDNEY.

SCALE.

KANGAROOS.

NATIVES OF AUSTRALIA FELIX.

The Illustrations by J. Marchant, & Engraved by J.B. Allen.

The Map Drawn & Engraved by J. Rapkin.

JOHN TALLIS & COMPANY, LONDON & NEW YORK.

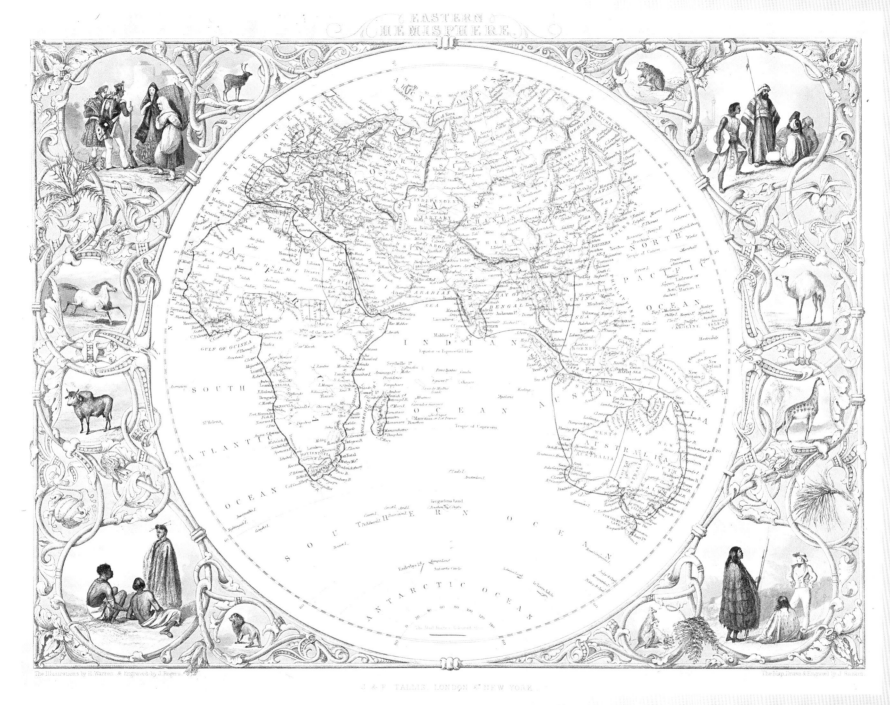

Eastern Hemisphere

1851 **Tallis**

Featuring the imperial mail route from Great Britain eastwards to India, Singapore, Hong Kong and Australia. Four vignettes represent people of different racial origins from throughout the hemisphere. This map with Rapkin's decorative vignettes appeared in Tallis's *Illustrated Atlas* in 1851.

Post-Settlement Van Diemen's Land

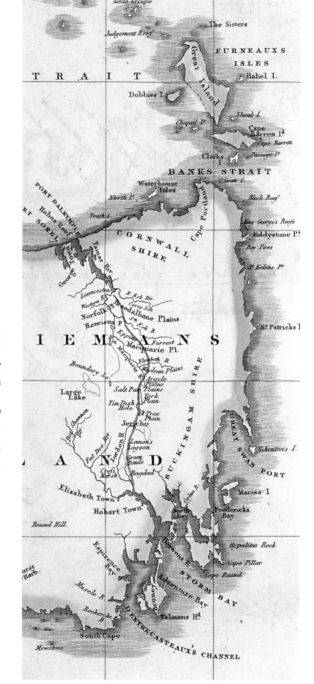

Map of Van Diemans Land and
Map of Part of New South Wales

1824 **Lizars**[23]

The Hobart-Launceston road opened up the Midlands to
European settlement.

Detail
Vignette of Hobart Town.

HOBART TOWN, VAN DIEMANS LAND

This series of maps, from 1824 to c. 1860, records the
spread of settlement throughout the island. In January
1856 Van Diemen's Land was renamed Tasmania, and
in December became a self-governing colony.

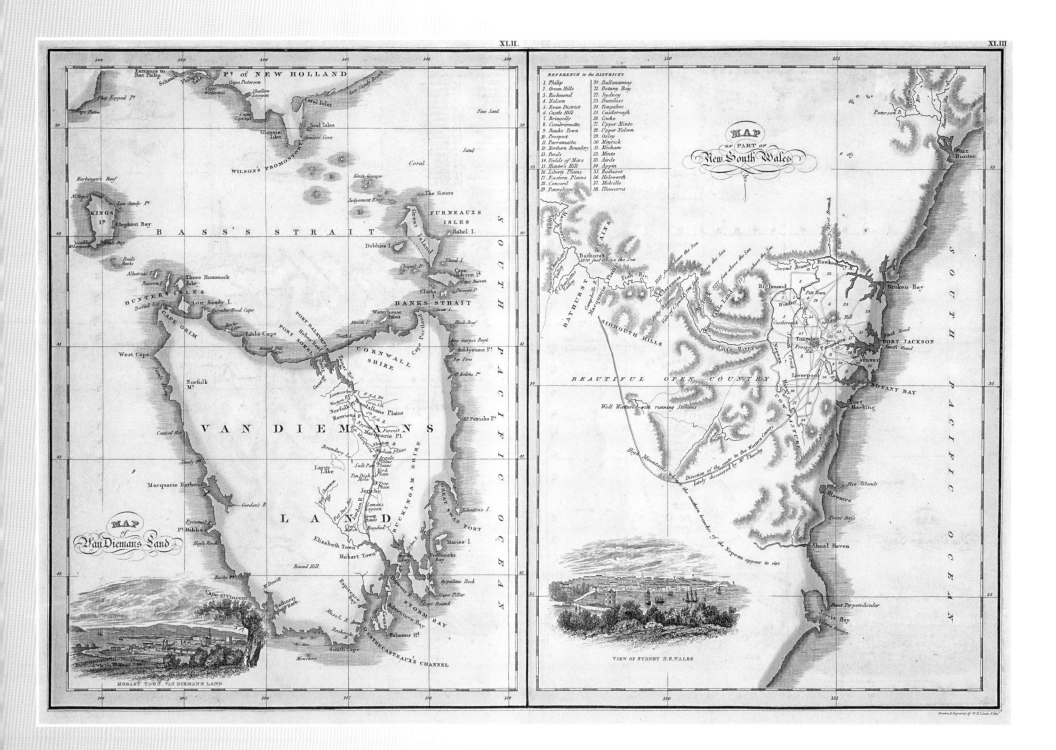

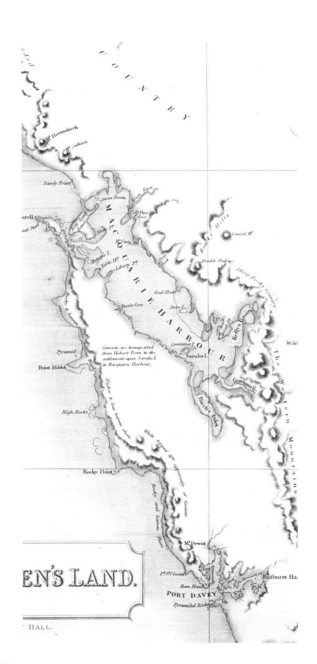

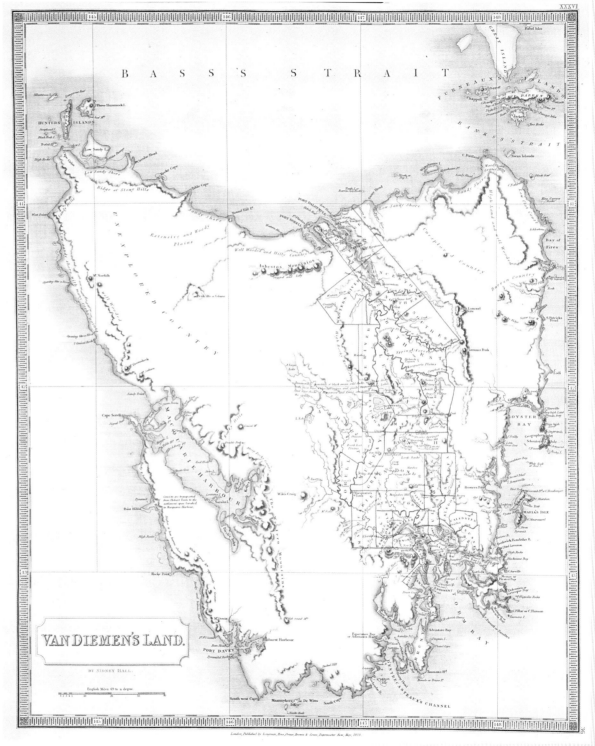

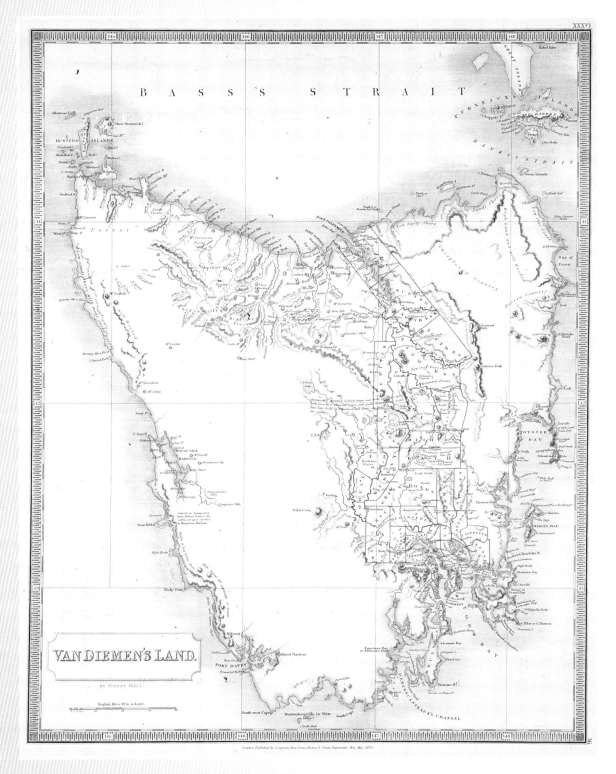

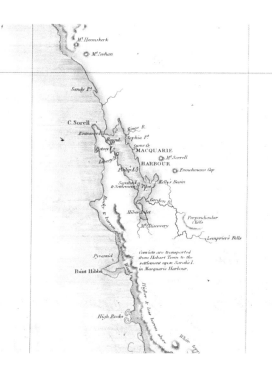

Van Diemen's Land

TWO MAPS

1828 Hall[24]

Both with the same title and date, these two maps illustrate an obvious cartographic error, with one depicting Macquarie Harbour as twice its actual size. The first map was quickly withdrawn, the error corrected, and the map republished. Additional topographical information also appears in the later map. The isolation of Macquarie Harbour, in which the grim penal settlement of Sarah Island is located, is clearly evident. Published in Hall's *New General Atlas*.

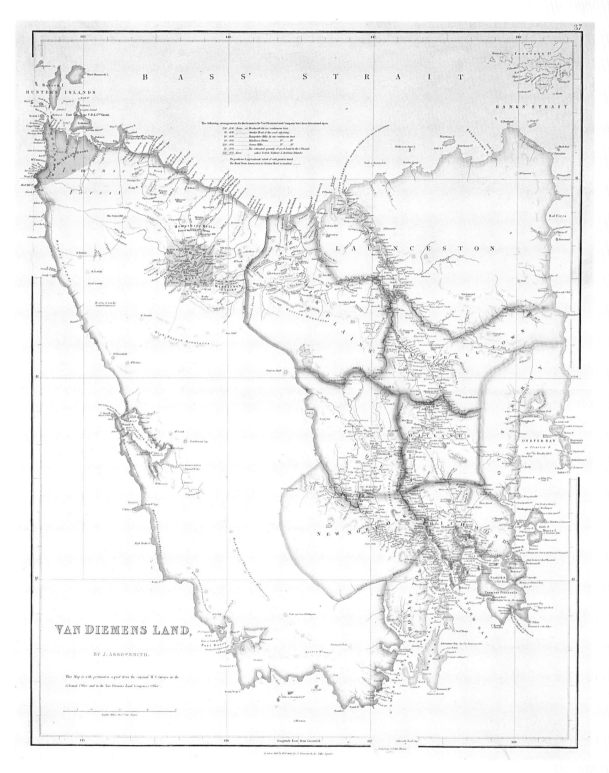

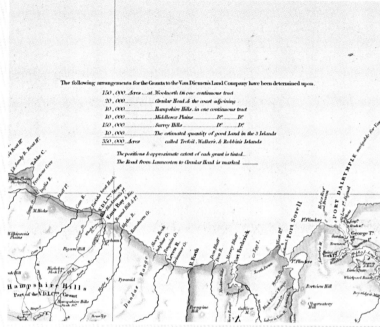

Van Diemens Land

1842 J. Arrowsmith

The solid pink coloured areas to the north-west of the island, indicate the 350,000 acres [141,000 hectares] granted to the Van Diemen's Land Company.
Included in the *London Atlas of Universal Geography*, 1848.

Detail

Explanatory note detailing the various V.D.L. Company grants.

Van Diemen's Island or Tasmania

1851 **Tallis**

This decorative map includes delicate engravings of the now extinct thylacine or Tasmanian tiger (*Thylacinus cynocephalus*) and the 'Residence of the V.D.L. Company's Agent, Circular Head'.

Details

A mid-19th century view of the thylacine, depicted erroneously as having no stripes.

Vignette of Hobart Town.

HOBART TOWN.

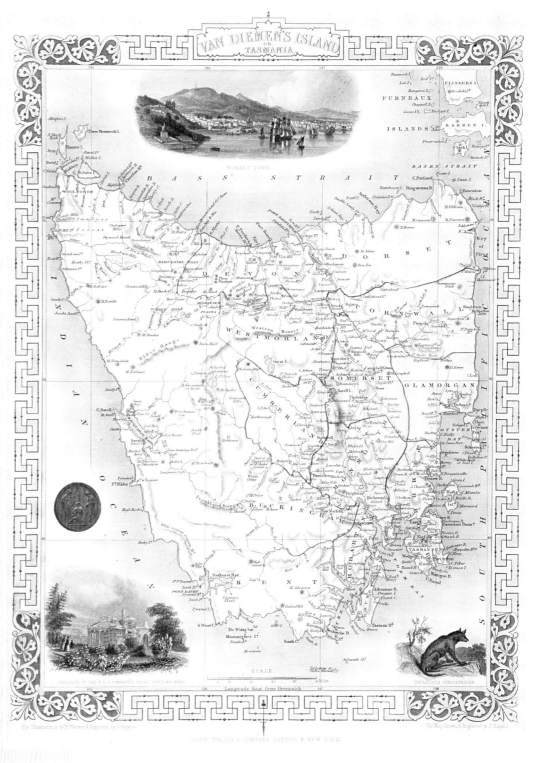

To early geographers and navigators, *Terra Australis Incognita* was a hypothetical landmass at the bottom of the globe. It was the prospect of wealth and riches, together with trading rivalry, that prompted nations to send expeditions in search of this mythical land. With the Dutch established in the spice-rich East Indies (now Indonesia), explorers were sent to probe south. But as the outline of the continent of Australia began to emerge, there remained a belief that this southern continent may not be the fabled *Terra Australis Incognita*. As a result, Tasman, while endeavouring to establish whether or not there was another southern landmass, encountered Tasmania. However, it was Cook who delineated the eastern extent of Australia and, during his second voyage in which he sailed into Antarctica, proved that there was no other continent which could claim to be *Terra Australis Incognita*. It remained only for Bass and Flinders to circumnavigate Tasmania for the full outline of the island to finally emerge.

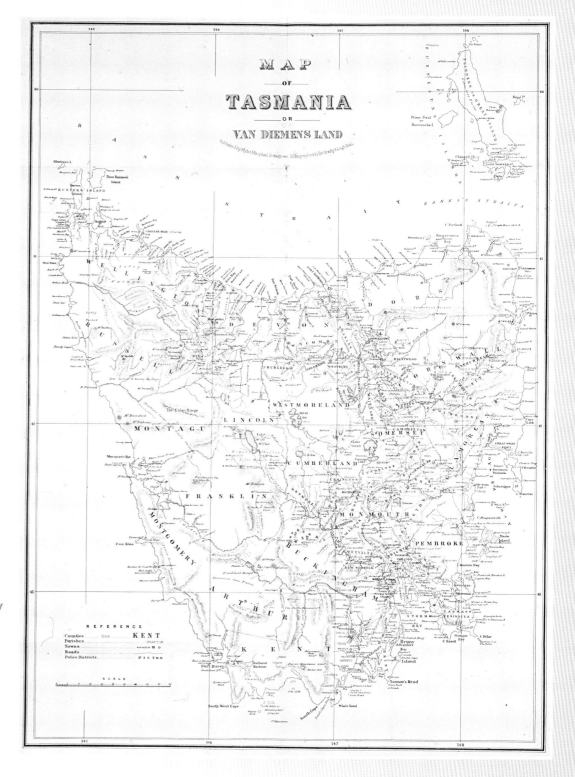

Map of Tasmania or Van Diemens Land

DATE UNKNOWN **De Gruchy & Leigh**[25]

In 1856, in an attempt to distance itself from its convict past, Van Diemen's Land was renamed Tasmania. For some years it was not uncommon for maps to record both names, such as in this early 1860s example.

Endnotes

1. **Abraham Ortelius**, geographer and cartographer, was born in Antwerp in 1527. In 1570 he produced his most ambitious work, *Theatrum Orbis Terrarum* (Atlas of the Whole World). This was a collection of contemporary maps, which he engraved to a uniform size. It is believed to be the first such compilation.

2. **Henricus Hondius**, son of noted cartographer Jodocus Hondius, worked in the thriving family business. The Hondius cartographic and publishing firm was one of the most prominent in Amsterdam. During the 1630s, Hondius's brother-in-law, Johannes Jansson, joined the firm as a partner.

3. **Johannes Jansson** was married to Elisabeth, a member of the great Hondius cartographic family. He became a partner in the family business early in the 1630s, and, after the death of his brother-in-law, Henricus Hondius, Jansson took over the firm. He died in 1664.

4. **Pieter Goos**, a prominent Amsterdam cartographer, expanded the business begun by his engraver father, Abraham. Goos was well known for his sea charts. He died in 1675.

5. **Frederick de Wit**, born in 1630, became a formidable cartographer and publisher in the latter part of the 17th century. One of the last great Dutch mapmakers, after his death in 1706 all his plates were sold and his publishing house ceased to exist.

6. **Hugo** and **Caral Allardt** were Amsterdam mapmakers. Hugo Allardt founded the business, which was continued by his son Caral on his death in 1691. The Allardts generally included a number of maps from other cartographers in their publications; consequently, their own individual maps are quite rare.

7. **Guillaume de l'Isle**, the leading French cartographer of the early 1700s, was prominent when the highly decorative school of Dutch cartography was superseded by a more scientific approach. De l'Isle set a new standard of accuracy, using up-to-date celestial observations in calculating latitude and longitude. He was appointed as France's first Royal Geographer in recognition of his skills. After his death in 1726, a number of his copperplates were acquired, updated and reissued in Amsterdam by Covens & Mortier, prominent 18th century Dutch cartographers and publishers.

8. **Gilles** and **Didier Robert de Vaugondy** were French cartographers and publishers, noted for their intricate illustrations and cartouches. Didier succeeded his father Gilles, but after Didier's death in 1786, the firm passed to cartographer Charles Delamarche.

9. **Robert Benard**, a notable French engraver and mapmaker, reproduced many of the charts for the French editions of 'Cook's Voyage'.

10. **Rigobert Bonne** was a well-respected French cartographer who produced high quality maps. Appointed Hydrographer to the Navy, he had access to the latest French survey information. He died in 1795.

11. **Daniel Djurberg**, a Swedish publisher and geographer, was the first mapmaker to adopt the Maori word 'Ulimaroa' to describe Australia. He died in 1834.

12. **Johann Weigel** and **Adam Gottlieb Schneider** were cartographers and publishers who worked in Nuremberg. 'Schneiderschen' has been incorrectly translated in the past as Schneider's name. The map itself is Weigel and Schneider's work.

13. **Giovanni Maria Cassini**, a major Italian mapmaker, engraver and artist, spent five years collecting and engraving the maps for his popular 1801 atlas, *Nuovo Atlante Geografico Universale*. He died in 1824.

14. **Charles-François Beautemps-Beaupré** accompanied Bruni d'Entrecasteaux when he visited Van Diemen's Land in 1792 and 1793. Meticulous and precise, and a highly respected hydrographic engineer and cartographer, Beautemps-Beaupré was a member of the Academy of Sciences.

15. **John Hayes** (later Sir John), a naval officer and explorer, served in the Bombay Marine of the East India Company. He was awarded a knighthood in recognition of his fine war record, and died in 1831, aged 63, in the Cocos Islands.

16. **Aaron Arrowsmith** established his firm in 1790. By the early 19th century, the Arrowsmiths were the leading British map publishers, providing the most reliable records available at the time. After his death in 1833 the business was carried on by his two sons and then by his nephew, John Arrowsmith, who became the most famous member of the family.

17. **Johann Mathias Christoph Reinecke** was a cartographer for the Geographical Institute in Weimar. He died in 1818.

18. **Pierre Lapie** was appointed First Geographer to the King of France in the post-Napoleonic War period. He was also a military cartographer.

19. **Victor Levasseur**, a French geographer and publisher, produced his popular *L'Atlas National Illustré* in 1838. Admired for its artistic presentation, his work often featured beautiful illustrations and etchings by Raimond Bonheur and Frederick Laguillermie.

20. **Society for the Diffusion of Useful Knowledge** was founded in 1826, with the aim of producing relatively inexpensive maps to encourage broader education in Great Britain.

21. In 1810 **John Arrowsmith** succeeded his uncle and cousins in their successful family business. He was instrumental in publishing maps of inland Australia. The Arrowsmiths produced more than 70 maps of the Australian continent, including revisions. He died in 1873.

22. **John Tallis**, a 19th century cartographer, founded the London publishing house of Tallis & Company. Tallis's 1851 *Illustrated Atlas*, which featured extensive use of vignettes, was one of the last decorative atlases.

23. **William Home Lizars** spent some time as an artist before joining the family business in 1812. His artistic ability is revealed in his finely etched vignettes.

24. **Sidney Hall** was a respected 19th century London engraver who worked in the Strand, then in Bloomsbury.

25. **De Gruchy & Leigh** were lithographers with a business in Elizabeth Street, Melbourne.

About the Authors

Authors **Libby** and **John McMahon** have a passion for history. Libby graduated in Arts from the University of New England, majoring in archaeology. Her fascination with old maps increased after she and her husband spent two years navigating and sailing their yacht off the coast of northern Queensland.

John, as a regular army officer, saw active service during the Malayan Emergency and in Vietnam. He has undertaken research into early colonial military history and been awarded the degrees of Master of Humanities and Doctor of Philosophy from the University of Tasmania.

Further Reading

Bricker, C.
Landmarks of Mapmaking:
an Illustrated Survey of Maps and Mapmaking
Phaidon, Oxford, 1976.
Reprinted Wordsworth Editions, Hertfordshire, 1989.

Clancy, R.
The Mapping of Terra Australis
Universal Press, Macquarie Park, 1995.

Clancy, R. & Richardson, A.
So Came They South
Shakespeare Head Press, Sydney, 1988.

Pearson, M.
Great Southern Land:
the Maritime Exploration of Terra Australis
Dept of the Environment and Heritage, Canberra, 2005.

Sharp A.
The Discovery of Australia
Oxford University Press, London, 1963.

Tooley, R.V.
Printed Maps of Tasmania 1642—1900
Francis Edwards, London, 1975.

Tooley, R.V.
The Mapping of Australia and Antarctica
2nd edn, New Holland Press, London, 1985.